FOOD SERVICE
Modern Techniques & Practices

FOOD SERVICE
Modern Techniques & Practices

Edited by
Jyoti S. Sharma

AKANSHA PUBLISHING HOUSE
NEW DELHI - 110 002 (INDIA)

AKANSHA PUBLISHING HOUSE
4649/21-B, Ansari Road
Daryaganj, New Delhi - 110 002
Phones: 23263193, 9811582579
E-mail: ektabooks@yahoo.com

Food Service: Modern Techniques & Practices

First Published, 2006

ISBN 81-8370-068-3

PRINTED IN INDIA

Published by M.P. Misra for Akansha Publishing House, New Delhi
and Printed at Aryan Enterprises, Delhi

Contents

Preface

The foodservice (or food service) industry generally encompasses those places, institutions, and companies responsible for any meal eaten away from home. This industry includes restaurants, school and hospital cafeterias, catering operations, and many other formats. Major foodservice providers include Compass Group, Sodexho, Aramark, and the Crown Group.

The companies that supply foodservice operators are called foodservice distributors. Some companies manufacture products in both consumer and foodservice versions. The consumer version usually comes in individual-sized packages with elaborate label design for retail sale. The foodservice version is packaged in a much larger industrial size and often lacks the colorful label designs of the consumer version.

In fact, food service is a core area of hotel and hospitality industry. As food is the most basic constituent of human survival on the earth, food service is of prime importance. By food service, here we don't mean only the currently or past traditional eliquettes of serving the food to customers/ guests, but entire process of food preparation, its conservation and presentation.

In this book, we have compiled valuable information of food service, its various relevant facets and techniques. The information is compiled under the following headings, namely—Fundamentals of Food Preparation; Method of Food Preparation; Fundamentals of Menu Planning; Cooking for Patients; Sea Foods; Dishes of Eggs; Cereals and Pulses;

Dishes of Meat; Poultry Dishes; Vegetable Dishes; and General Guidelines etc.

An exhaustic bibliography, glossary of terms encompassing various aspects, etc. are added features of the book. Definitely amidst the plenty of literature, this book will serve as a vade mecum to learners, trainers and professionals in the field.

The entire information is gathered from various authoritative secondary sources, e.g., books, menus, abstracts, catalogues etc. The leading experts and professional's views and experiences are heavily drawn upon for the information. We owe a deep sense of gratitude to them all.

We are thankful to our friends, colleagues and well wishers for extending ready help, assistance and also enthusing our present efforts. Finally, we are thankful to our publishers for bringing out this book superbly within stimulated time period.

Constructive criticism and suggestions are welcome from all quarters for enhancing the quality and utility of this work.

—Editor

1

Fundamentals of Food Preparation

Your forefathers did without sugar until the 13th century, without coal fires until the 14th century, without buttered bread until the 15th century, without potatoes until the 16th century, without coffee, tea, and soap until the 17th century, without pudding until the 18th century, without gas, matches and electricity until the 19th century and without canned goods until the 20th century.

Cooking aims to facilitate and hasten digestion, so that the cooked food is absorbed by the digestive system and subsequently assimilated by the body. This is largely determined in the manner the food is cooked. During the cooking process, it breaks down the cellulose in plant food, softens some of the connective tissues of meat, breaks down and gets starches present. The alteration is brought about in texture, by physical and chemical changes thus assisting mastication.

A physical change occurs when a substance changes its form, colour or size, but still remains that same substance, like water that changes to ice. A chemical change occurs when a substance changes its form, colour or size, combining so as to form an entirely new body, e.g. green marrow changes its colour and milk changes to curd.

The effect of cooking upon the three chief constituents of food-- proteins, carbohydrates and fats--is visible in their increased digestibility.

Proteins

The protein of meat (myosin), of egg (albumen), of wheat (gluten), of pulse (legumin) is coagulated by heat. Avoid high temperatures as the protein hardens, denatures and shrinks and the food becomes indigestible. The connective tissue is converted into gelatine which is soluble in water and rendered digestible. The biological value of proteins is improved by moderate heating.

Carbohydrates

(1) Starch in food is greatly affected by heat. By moist heat, it is converted first into a soluble form and then by extreme heat into a new substance, sweetish in flavour--dextrin as in the crust of bread. Moist heat causes the starch grains to swell; it gelatinises at a temperature below boiling point of water, the degree of heat varying with the kind of starchy food.

(2) Cellulose is softened by the application of moist heat.

(3) On heating, sugar in water dissolves, then colours, upon further heating, turns brown and becomes a caramel and emits a lovely flavour, but does not crystallise.

Moderate heat does not cause much loss of mineral salts and vitamins, Vitamin C being.

Fats

(1) If heated to a very high degree for a long time, fats undergo partial decomposition and fatty acids and glycerol are produced. Glycerol further decomposes into acerolin which is an irritating compound to the digestive system.

(2) Cooking pleases the eye and is receptive to the palate and helps to stimulate the digestive juices,

thereby creating an appetite. On application of heat, the red colouring matter such as haemoglobin in meat decomposes and changes the red colour to brown. Flavours are developed, which simulate the digestive juices. The colours of the vegetables are brightened. Green vegetables fast boiled without a lid, improve the green colour. Cauliflower when boiled, in acidic medium, gets a white refreshed look.

(3) Cooking sterilizes the food partially. Cooked food can be stored for a longer time and it prevents food poisoning and diseases when stored properly. Some of the disease producing germs are killed by cooking. They are killed because of high temperature during the cooking process. A temperature of 60° (140°F) applied over 30 or more minutes, kills most of the pathogenic germs. However, there are some spore germs which are not destroyed by the usual cooking temperature: they take about 4 to 5 hours to get destroyed at a high temperature. High temperature penetrates to the centre or interior of food, specially in pork, beef and mutton.

(4) Cooking preserves food for a longer time. The high temperature destroys bacteria and limits spoilage. It is economical as the cooked leftovers could be utilised and interesting new dishes could be prepared.

(5) Cooking retains, as far as possible, the nutritive and flavouring ingredients. The flavour depends upon the amount and kind of extractives present, and the acids developed. Nutritive value is enhanced if the fat proportion in the meat is more. While cooking, the nutrition could be preserved by using the cooking liquor.

(6) Cooking gives a variety to the menu, as one food item could be cooked in various ways and given different textures, e.g. mutton in a soup, roast joint,

croquettes, stews, steaks, keema, sookha meat, boti kababs, etc. Different methods of cooking when used, make the menu interesting and enhance variety. It is, therefore, easier to plan balanced diet.

Raw Material Used: For the preparation of good, tasty, colourful dishes, it is essential to have a basic knowledge of the raw materials, their characteristics and the special part they play. This knowledge helps to substitute materials when necessary, or rectify the texture and taste if something goes wrong. Also it helps to improve the quality and get standard end product.

The raw materials are thus classified according to the part they play in making up a dish:

(1) Salt
(2) Liquids
(3) Sweetenings
(4) Fats and oils
(5) Raising agents
(6) Thickenings
(7) Flavourings and seasonings
(8) Eggs

Salt: It brings out the flavour of other ingredients. The other name for salt is sodium chloride: it is readily available in all parts of the world in a solid (rock salt) or in a solution form (sea salt). Salt, having a distinctive taste, transforms an insipid dish to a wonderful dish. It should be used skilfully, or too much of it could spoil the dish.

Generally, salt is available in three forms: 1. Table salt (fine) containing phosphate; (2) Coarse or freezing salt for culinary purposes; (3) Celery salt. It is a blend of celery root and ordinary salt and is purchased ready prepared. It is used for flavouring certain dishes as an alternative to fresh celery or celery seed.

Various Uses

(1) It has a physical effect on the gluten of flour and strengthens gluten and increases its resistance to the softening effects of fermentation.

(2) Use of the correct amount of salt improves the flavour of the savoury dishes and when a little is added to sweet dishes, it enhances the flavour.

(3) Cauliflower, when put in salted water, makes the insects come out.

(4) It has a controlling effect on the activity of yeast in bakery products. It controls fermentation and hence it has marked effect on crumb, crust and colour of baked products.

(5) Salt added to water, for cooking green vegetables, helps in colour retention and enhances the taste.

(6) Salt is essential for good health.

(7) Salting is one of the oldest popular methods of preserving ham, bacon, fish etc.

Liquids: Liquids are important as they are used for the purpose of cooking, binding and coating, etc. Milk, water, stock and fruit juices are the most commonly used liquids. They prevent food from burning, bind dry ingredients and help in maintaining the right consistency.

Water and milk are used for preparing poaching liquor *[court bouillon)* soups, sauces, gravies, cakes and pastry mixtures and kneading of doughs, etc.

Buttermilk is used for curries, kadi, etc.

Stock is a liquid containing soluble nutrients and flavours, which are extracted by prolonged and gentle simmering (except fish stock-- 20 mins.). They are used as the foundation of soups, sauces, gravies, curries, poaching liquors and many important kitchen preparations.

It is important that the correct amount of liquid should he used, or too much would make the food stodgy, soggy or watery.

Sweetening

When sweetening is used with other foods it enhances the combined sensations of odour and flavour of the dish produced. It also adds its own sweetness, and is a versatile food product. Its uses in-the kitchen are varied. Substitution of one sugar (castor, icing) for another in a baking formula, gives allowance for the difference in the sweetening effect. Sweetening is available in various forms--granulated, fine-grained, powdered and in a solution form.

Sugars vary in their sweetening quality and are available in the following forms-- granulated sugar, castor sugar, icing sugar, lactose (milk sugar), cane syrup, maple syrup, treacle, honey and golden syrup.

Functions of Sweetening

(1) The ability of sugar to crystallise, gives a delightful variety in cookery.

(2) Low concentration of sugar speeds the effectiveness of baker's yeast by providing an immediate, fast cooking course of nourishment for its growth--thus hastening the leavening process.

(3) Sugar gives puddings, breads, buns and bread rolls a good flavour (caramel), the characteristic golden brown colour, and a tender and even texture.

(4) Fermentation is aided by sugar in brewing, wine-making and in the distillation of spirits.

(5) Sugar has long been used as a preservative, it is an effective preservative and can dissolve in water at very high concentration.

(6) It is an energy food and can be stored for a long time without getting spoilt.

Bakery Functions

Stock Sugar Syrup

500 gm Cube sugar	120 gm
Glucose	1 litre Water

Boil all the ingredients together in a clean Sugar boiler. Strain through a cloth into a basin and keep covered.

Method and Temperatures of Cooking: Take the required amount from the Stock Sugar and boil to the correct temperature between 116° and 160°F.

Keep the side of the pot clean during the boiling, by washing the sides frequently with water.

Do Not Stir: Add a few drops of lemon juice and shake well.

Pulled Sugar: Pour on to a slightly oiled marble slab, allow to cool slightly, with a knife turn the sugar overfolding together, work the sugar with the hand to a pliable consistency.

Aniline powder is best for this type of work, being very strong in colour and easily dissolved in warm water.

The colour is added to the cooked sugar when poured onto the marble slab and before working with hands.

Degrees and Stages of Cooking the Sugar

116°C	Soft Ball	Suitable for Marzipan
119°C	Ball	Suitable for Fondant
121 °C	Hard Ball	Suitable for Nougat
140°C	Small Crack	Suitable for Italian Meringue
153°C	Crack	Suitable for Dipping Fruit
160°C	Hard Crack	Suitable for Pulled Sugar

Using Fats and Oils

Fats and oils are nutritionally useful and in some form, economical sources of energy and give a satiety value to the dish. They also contribute characteristic palatability, qualities of flavour and texture. They are popularly used as the medium of cooking.

Fats are solid at ordinary temperature and melt when heated. Oils are liquids at ordinary temperature. Only coconut oil solidifies at low temperature.

Various fats used in cooking are--lard (pig's fat), suet (fat around kidneys), dripping (Tallow Beef fat), butter, margarine, ghee, hydrogenated fat, cocoa butter (for confectionery).

Oils are extracted from coconut, palm, sesame, cotton seed, olive, peanut, mustard, corn and sunflower. Salad oil is a deodorised vegetable oil and is used for salad dressings, etc. as olive oil--the best for the purpose--is very expensive and scarce.

Fats and oils are used for various purposes, the major culinary part played by them are as--(1) spreads, (2) shortenings, (3) salad dressings, (4) frying media and (5) tempering.

Spreads: Butter and margarines are used for spreads, and their function is to add to the flavour, nutritional value and satiety value of breads.

Tempering: Dais, curries, rice dishes, etc. are tempered. The fat or oil is heated to which cumin seeds or mustard, or fenugreek seeds, etc. is added and poured over the dals.

Shortening: These are fats which shorten the gluten strands, surround them and make them more easily broken (short). When added to bread, it gives a bit of tenderness, richness and a sheen to the crumb.

Salad Dressings: Fat is used for the various salad dressings--Hot animal fat dressings, which consist of bacon fat, vinegar and seasonings, served hot, are used on green hot salads. Cooked dressing is a cooked mixture of egg, vinegar, fat, starch and seasonings. French dressings are emulsions of oil, vinegar or lemon juice and seasoning. Mayonnaise is an emulsion of oil, acid, egg yolk and seasoning.

Frying Medium: Fats and oils are used as a medium of cooking, i.e. pan roasting, frying, and sauteing. When fats or oils are heated, a temperature is reached at which visible fumes appear, which is defined as smoke point. Fats with high smoke point are suitable for frying. Different fats do not have the same smoke point. The highest frying needed for any food is about 199°C (390°C). Hydrogenated fats have a high smoke point and are good as a frying medium compared to other fats. For the selection of a good fat, the following points should be remembered.

A frying fat should have a high smoke point, low congealing point, high stability, low moisture content, and should have an acceptable flavour. Fats and oils have a high percentage of unsaturated fats and have more shortening power, than saturated fats. Animals fats such as butter, lard and suet make the food short, i.e. break off (short) and readily melt in the mouth.

For the selection of shortening, the following factors should be observed:

(1) Shortening should cream well or it will affect the cake volume.
(2) It should have no moisture; then it will be short. Shortness is correlated to baking biscuits, cookies, pastries, wafers, mathies, nankhatais, nimkies etc.
(3) The shortening should be stable and should not get rancid so that the cooked product has a good keeping quality.

(4) The right consistency of fat, i.e. solidified for puff and flaky pastry, soft for cakes and biscuits help to get a good texture. To acquire good results in baking, because of fats and other ingredients, certain rules should be followed-- The richer the pastry, the hotter the oven, and the richer the cake, the cooler the oven."

Clarification of Fat: Used fat should be clarified and then used for better results in cooking. Strain the used fat and then mix double the quantity of water in a pan and bring it to the boil. Strain again, cool and place it in a refrigerator. The fat will solidify and float on top. Lift the cake of fat, turn it upside down and scrape off the foreign particles that have collected. Heat the fat on slow fire, till the water evaporates and then strain and store it in a cool place.

Rendering of Fat: Animal fat is heated and melted and this renders fat from fatty tissues. Tallow, suet and lard are usually rendered and used for cooking. The fat is cut into small pieces and placed in a pan and put in the oven or on slow fire, until the fat melts, and there are crisp brown pieces of tissues left. This should be strained through a fine cloth into a clean bowl.

Leavening Agents: Leavening is increasing the surface area of a dough or batter by creating within myriads of gas bubbles puffing up, thus increasing the volume and making it light.

The expansion of these gases during baking increases the volume of the product and gives a desirable porous structure. The aeration of flour products is affected by the following:

(1) Biological (yeast)

(2) Chemical (baking powder)

(3) Mechanical (whisking, beating)
(4) Lamination (folding, rolling)
(5) Combination of the above.

Yeast

It is a living micro-organism and is a form of plant life. It requires for its growth, food, moisture, warmth and air. The primary function of yeast is to change sugar into carbon dioxide gas, so that the dough, in which it is generated, is aerated. It also assists in mellowing and ripening the gluten of the dough, and it contains vitamin B complex.

When yeast is dispersed in water at a suitable temperature and mixed to a smooth paste with flour and all the food necessary for fermentation are present, e.g. sugar, skimmed milk, the sugar acts as food and soluble proteins to form the structure and building material of the new cells. Activity starts within the yeast cell, yeast exudes a substance known as an enzyme which changes a solution of sugar (sucrose) and water into a simple sugar (dextrose) thus absorbed within the yeast cell and changed to carbon dioxide and alcohol, together with other by-products.

Other enzymes in flour and yeast change some of the soluble starch to sugar, which in turn is assimilated by the yeast and changed, so that aeration goes on from this continued production of gas.

When using yeast, these points should be remembered:

(a) It remains dormant in cold.
(b) It rises in warmth.
(c) It is killed by heat (if temperature is over 127° F).
(d) It is fed by sugar, and in the dough it caramelizes on the outside, forming a brown crust.

(e) It is fed by flour, the starch gelatinizes with the moisture inside the dough and cooks into a light spongy mixture which is digestible.

(f) Strong or hard flour is the best to use in bread-making.

(g) If salt is added in the correct proportions, it gives a good flavour, controls fermentation and improves the colour of the finished goods.

Commercial yeast is of two main types compressed in cake form (also in a liquid form) and dry yeast in granular form.

Compressed yeast is a moist mixture of yeast plants and starch. The yeast remains active and multiplies rapidly if added to the dough. It should be kept in the deep freezer, because it retains its activity for several weeks, or even months. Liquid yeasts and starters are sometimes used in household bread-making. They are mixtures of active yeasts from compressed form with yeast food, such as sugar and potato flour or dough. They have a short storage life.

(i) Dry yeast: It is a mixture of yeast with corn meal or starch pressed into cakes and dried. Such yeasts continue to live, but are in an inactive state. When furnished with food and moisture they begin to develop and multiply, but do so slowly. They can be stored for a long time and are inexpensive.

(ii) Activated dry yeast: It develops faster than dry yeast and is less perishable than compressed yeast. It is used in straight dough mixing. It should be stored at refrigerator temperature for a longer life.

Chemicals: Chemical aeration is brought about by the production of carbon dioxide from the action, in solution, of an alkali and an acid in the presence of heat. The acid and alkali in correct proportions (which is harmless to human digestion) is baking powder. With this method of aeration, the greater amount of gas should be generated after heat has been applied, i.e. when the goods are in the oven.

The carbon dioxide which is generated passes into the air cells, already created and is held by the network of gluten in the dough. This generation of gas and the consequent expansion causes an increase in the volume, which is held by the coagulation of the gluten and other proteins that are present.

The dough cooked becomes light and digestible. Baking powder consists of two parts acid and one part alkali. This can be bought readymade or it can be made as per the following recipes:

Recipe I:	Cream of tartar	60 gms
	Bicarbonate of soda	32 gms

Corn-flour (optional, increases the stability of the mixture.)

Method: Sieve together several times. It has to be kept in an airtight tin. The addition of one part rice flour or Corn-flour will help to keep dry.

Recipe II:	Tartaric acid	30 gms
	Bicarbonate	30 gms

Method: Same as above.

The latter recipe is suitable, but once mojsteried it reacts rapidly, so that much of the gas is lost, if delayed in reaching the oven.

Other acids which may be used, when combined with bicarbonate of soda will produce carbon dioxide such as vinegar and sour milk. When sodium bicarbonate is heated, it gives off part of its carbon dioxide, but leaves sodium carbonate as a residue. This compound is unpalatable and hence should always be neutralized with an acid which leaves a relatively tasteless and harmless salt.

If no sodium bicarbonate is added, the cookies are pale in colour; when it is added, the intensity of colour increases with the quantity of soda. Cream of tartar used in angel cakes increases their activity and tends to make them white and fine grained.

Mechanical Aeration: It is incorporating air by whisking, beating and sieving. When sugar and eggs, fat and sugar, fat and flour or any combination of these are beaten or whisked together, it works as aeration. Whichever way the air is introduced into the mixing, be it by hand, whisk, spatula or by machine, it is still termed mechanical aeration. Air is incorporated into the mixing by one of the above means and is held there by the fat, eggs, or both; aeration is brought about by the expansion of the air in the hot oven, together with the water vapour pressure within the air bubbles.

One of the best examples of mechanical aeration is the sponge cake, where a foam is produced from the eggs and sugar, the flour is then folded in the sponge and baked. The egg and flour proteins coagulate and the starch cooks thereby making the sponge set.

Water Vapour: Lamination acts as a raising agent. In pastries-- folding and rolling helps to give the lift. In idlis, khaman dhoklas, etc. steam formed by the heat, helps to puff up, as liquid and flour are present in equal quantities. As the popcorns have moisture inside the grains, when heated, they expand in volume. The fundamental objective when making puff pastry is to build up a structure of fat and dough consisting of many layers so that when subjected to heat in an oven, it will expand and lift evenly, to produce goods with short eating properties.

In the oven, the pastry, which consists of thin layers

of dough separated by films of fat, comes under the influence of heat, and the gluten in the dough layers is caused to expand and blister. The fat melts and the dough layers are insulated and the fat takes on a higher temperature, dough layers are cooked, the gluten coagulates and becomes almost rigid and the pastry does not collapse.

It is the expansion and blistering of the gluten in the dough layers as a result of steam pressure from the water in the dough that is responsible for the lift.

Combinations: Danish pastries are a combination of aeration by yeast and by lamination. For this, a rich yeast dough is made and butter is incorporated as for puff pastry. In this, the pastry expands in volume, because of the process of aeration, by lamination, and at the same time is aerated by the action of yeast.

Thickening Agents: Thickening agents give body, consistency, and palatability, when used. They improve the nutritive value. Flavoured liquids are thickened and converted into soups, sauces, gravies, curries, bavarois, mousses, puddings, etc. Binding agents are used to form a mixture of ingredients into a cohesive mass.

The thickening agents are starch, agar, eggs, gelatin, coconut, tamarind, curd, poppy seeds, onion paste, coriander power, etc. Starch is the reserve carbohydrate of plants and is abundant in common foods: corn, rice, wheat, potatoes, arrowroot, and tapioca; rice has about 80% starch, wheat 70% potatoes about 19% starch.

Cereal starches, root starches and instant starches are major varieties of starch.

Cereal Starch: Rice, wheat and maize are major sources of cereal starch. When cooked, corn, wheat, or rice, begins to gelatinize at about 71° (160°F) but requires about 85-96°C (185-205°F) for maximum thickening. Cereal starches when cooked and cooled tend to gel, while most root starches do not, as they have more thickening power than root starches. Root starches are tapioca, potato and arrowroot which thicken at a lower temperature of 65-71°C (150-160°F) than cereal starches. They thicken completely below boiling point. Instant starches are pre-cooked starches, also known as pre-gelatinised starches. Instant puddings, etc. in the market contain instant starches.

Agar is a rigid transparent gel, and is used for salads, sweet dishes. It is a seaweed, widely found in China.

Eggs give a firm gel when baked, and thicken, but are soft when stirred.

Gelatin (unflavoured) is used for salads, cold sweets, cold soups. When set, it is transparent, firm and quivery.

Coconut is used for curries, and gives the gravy a whitish colour, thickens and enhances the taste.

Tamarind is used in a pulpy form and it gives a thickening to the consistency of the gravy and gives a brownish appearance and an acidic taste.

Curd is used for thickening curries and it gives a whitish appearance, and a good flavour.

Poppy seeds, onion paste, coriander powder give body to the curries, and slightly thicken the gravy and blend with the flavour of the curry.

SPICES AND HERBS

Spices and herbs give flavouring and seasoning to the dishes. To get effective results, not only should the food please the eye, but should also flatter or stimulate the palate. The success of cooking largely depends upon the help we obtain from flavouring and seasoning. The spice we use for this purpose should be used sparingly, as well as with skill. All palates may not crave for highly spiced food, yet majority of people demand that the food be moderately flavoured with the right constituents. To use flavouring and seasoning rightly is a great accomplishment; the dish could be spoilt by being overseasoned. Seasoning should bring out the natural flavours of the main ingredients and blend with them.

Seasonings as such have little or no nutritive value but are valuable for they give variety to the dishes and have medicinal value.

Significance of Spices and Herbs

Help in Digestion: From prehistoric times, spices have been used. Clove oil stimulates the flow of gastric juices; garlic, aniseed and asafoetida are taken for better digestion, and to reduce the chances of hypertension.

Medicinal Use: For toothache, clove oil relieves pain. Turmeric and oil applied on swellings and hurts, cures, as it is believed to have antisceptic qualities, etc. Garlic and saunf help digestion, ginger added to tea helps cure colds. Turmeric added to milk is given to a person who is in a state of shock.

For Flavour: Dishes would be insipid and bland if the spices were not added, because they give a good flavour and stimulate appetite. The flavourings added should not be overpowering, but there should not be a suspicion that they have been added. This mental search for the flavour makes the dish more interesting. Monosodium glutamate is a flavour enhancer in meat and fish dishes. Cinnamon, bay leaf, nutmeg, saffron, pepper, cloves, etc. also enhance flavour.

Monosodium glutamate is salt like crystals known as MSG. It is marketed under various brand names and is produced by a special process from wheat, soya and sugar beet. It enhances flavour in the cooking of savoury dishes.

For Nice Appearance: Some of the spices give colour to the food and improve the appearance of the dish. Turmeric, saffron, coriander leaves, poppy seeds, ratanjot and colour to the food, which makes the dish interesting.

For Palatability: Salt is one of the important seasonings that enhances the taste of the food. It also brings out the flavour. The other seasonings that improve the palatability are pepper, chillies, poppy seeds, coriander seeds and paprika, etc.

As Preservatives: Many food are preserved for a longer time with the help of spices. Salt is used extensively for preserving--brine solution. Other spices that have the preservative quality are-- turmeric, cloves, mustard, ginger, garlic, asafoetida. Pickling is one of the forms of preserving.

Spices that have flavourings and seasonings are: Garlic, ginger, cloves, cinnamon, cardamom, cumin seeds, mustard seeds, poppy seeds, nutmeg, asafoetida, coriander powder, mace, pepper, fenugreek, chillies, saffron, aniseeds, turmeric, paprika, caraway seeds, allspice, sesame.

Various herbs are parsley, celery, coriander leaves, thyme, tarragon, rosemary, mint, marjoram, sage, bayleaf, basil, chervil, caripatta, etc.

Spices and herbs are available in different forms: fresh, dehydrated and powdered, liquids and some are essences (concentrated) of spices, e.g. clove, cinnamon, etc. Many combinations of spices in a powdered form are marketed today, e.g. chana masala, rasam powder, sambar, masala for onions, garam masala, jal zeera masala, meat masala, etc.

This process retains original flavours and improves keeping quality and makes transportation economical and easy. Spices are also available in liquid form.

Convenience (Fast Food)

Today the term "Convenience Food" is widely used in the catering industry. This is probably due to the rapid progress in food technology over recent past.

Many convenience foods have been in use in the industry for many years, without being given this title. This was probably due to the fact, that they were used in small proportions during the preparation of meals. Considerable amounts of convenience food are now used in all levels of catering, packaged and graded specially for the industry.

This varies from individual types of food to complete meals. The term "Convenience Food", strictly translated, refers to any type of food, where some stage of preparation has been completed beforehand.

Processes: The following are some of the facts and methods used at present. Undoubtedly further progress in new developments will bring forth many more.

Accelerated Freeze Drying: By using high-vacuum techniques it is possible to produce specific conditions of temperature and pressure which create ideal conditions for the dehydration of food without the collapse of cells. Foodstuffs so treated have greatly improved texture retention and rehydration potentialities. There are only a limited number of food in the market, as production costs are relatively high.

Additives: Non-nutritive substances added to food (generally in small quantities) to improve its appearance, flavour, texture and storage properties. Most additives are legally controlled.

"Boil-in-the-bag" Ready Food: A convenience entree or vegetable commodity Portioned items are packaged in special plastic bags, sealed, then frozen. To reconstitute the

unopened bag is dropped into boiling water for a predetermined period.

Aseptic Canning: A newer development in canning designed for those products which are adversely affected by heat within sealed containers. Products like fruit juices are sterilized, then placed into sterile cans under microbe-free conditions.

Dehydration: A method of preserving food by reducing the moisture content to such a level that microbes cannot grow and reproduce. The methods used depend on the food but popular techniques include tunnel, roller and spray-drying.

Nutritionally the product lose little value, but the texture, especially in fruit and vegetable, is suspect. Newes methods (dehydro-freezing, freezing dehydration and quick dehydration) are overcoming this problem. A large variety of food are now available, e.g. pulses and leafy vegetables

Dehydro-freezing: A method of preservation where the moisture content of certain food is reduced by 70 per cent, then quick frozen. Because the cells do not collapse, the product has good texture retention. Product must be stored in a frozen state.

Freezer Burn: In any frozen storage room there is a steady removal of moisture from unprotected food which may result in unpleasant irreversible changes to the colour texture, flavour and nutritive values.

Freezing: Food freeze over a wide range of temperatures, although their individual freezing points are unidentifiable. Because of the high water content in many food, they freeze at temperatures between 32°F and 25°F. *Quick-freezing* has been defined as the process whereby food passes through this zone of maximum ice crystal formation (25-32°F) in 30 minutes or less. This ensures that minimum damage is done to the cell structure. Those caterers wishing to freeze their own produce should realise that slow freezing in deep-freeze cabinets might result in the production of large ice crystals, with subsequent breakdown of tissues, especially in fruits and vegetables.

Instant-freezing Preservation: Method of preservation using liquid nitrous oxide at 128°F below zero. By gently flushing food through nitrogen, freezing is almost instantaneous, producing very tiny ice crystals, so important for good texture retention.

"Free-flowing" Products: Many quick-freezing methods (e.g. fluidised bed and flash freezing) produce commodities that have no surface frost, and therefore do not adhere in the pack. Advantageous where thawed appearance is important, e.g. fruit.

Irradiation: A novel method of preservation, as it produces a raw sterile article with only a slight rise in temperature. The reaction depends on the absorption of energy by foodstuffs from a material or machine producing radiation. Process may be used for complete sterilization, pasteurization or as a pest and fungi controller, depending on the dose of radiation produced. Nutritionally there is little lost, and the dangers from radiation are negligible.

Preservation: Stored food tends to deteriorate under the influence of its own enzymes, chemical action and through the growth and reproduction of micro-organisms. The techniques developed for preservation aim to retard these reactions or destroy the cause of deterioration. Methods include: refrigeration, quick freezing, canning, sterilization, pasteurization, irradiation and the addition of various chemicals.

Sublimation: Under certain conditions of temperature and pressure it is possible for water to change from ice to vapour without becoming a liquid at any time. This allows food to be dehydrated without the collapse of tissues," given products with excellent texture and rehydration characteristics.

Spray Drying: A method of dehydrating food by spraying a slurry of the commodity into a hot-air chamber, which dries the material immediately forming a powder. Good rehydration properties and nutrient retention in milk, dried soups and instant coffees.

Smoked Preservation: This process is applied to a wide variety of raw and prepared foods. When used with fish, usually the fish is salt cured before smoking. Further examples relating to smoked sausages, meats, etc. need to be given.

Besides the different methods mentioned above which produce the greater part convenience food in dry, quick frozen and canned food, there are products which are very suitable for large-scale catering and produce quite an acceptable food item. When used with imagination and sometimes methods from conventional cookery, a much wider variety of dishes can be produced.

Ingredients Used: Various techniques are used for food preparation before cooking and they are done according to the requirements of the various dishes. This helps improve appearance, texture, palatability and flavour, and food combine readily. The techniques are divided into two:

(1) Sub-division and fractionalisation.

(2) Combining and mixing in the preparation of food.
Mise-en-Place: "Everything in its place", preparation prior to a task or service.

Sub-division and Fractionalisation

Washing: This is necessary to remove superficial dirt. Meat, fish, vegetables and fruits are washed in cold water before any preparation, i.e. peeling or cutting. If cut and soaked for a long period or washed after cutting, there is a great loss of water soluble vitamins and minerals. The more cut surfaces are exposed, the more nutrition is lost.

Paring: Paring is removing the surface layer in circular motion by pressure of a knife-edge all round the object.

Peeling: Spoilt, soiled and inedible portions are removed. Skin of vegetables or fruits is either peeled or scraped. The outer layer can be stripped by steam, or caustics such as lye solution.

Cutting: Reducing to small pieces with a knife. When a similar result is obtained with a chopping knife, or with a mechanical food chopper, the process is called chopping (small pieces).

Dicing--cutting into cubes is known as dicing, as in dicing potatoes, carrots, etc.

Mirepoix: Coarsely cut root vegetables. Desired shape for turning vegetable is barrel.

Mincing: Cutting into very fine pieces, e.g. mutton, onions, etc.

Meringue: Whipped egg white and sugar.

Mandolin: A kitchen utensil to cut vegetables and fruits to various shapes and sizes Julienne, Batons.

Macedoine: 6 to 8 mm cubes of vegetables or fruits.

Shredding: Cutting into long narrow pieces by means of a shredder or knife, e.g. cabbage.

Slicing: Cutting into thin pieces, but not so fine as shredding.

Slitting: Making a slit in the middle lengthwise, e.g. lady's fingers, green chillies, etc.

Grating: Reducing to small particles by rubbing on a rough surface, as in gratin lemon peels, cheese, etc.

Grinding: Reducing to small fragments by crushing, as in grinding spices, or coffee in a mill or on *a* grinding stone.

Mashing: This is a method of breaking up of soft food with pressure, with a potato masher or with a fork.

Pureeing: Food is strained and a Mouli should be used (veg masher).

Pressing: Separating liquid portions from solids by weights or mechanical pressure, as in making cider from apples, paneer, etc.

Sieving: Passing through a fine wire mesh to remove impurities. It also helps in enclosing air and mixing ingredients evenly, like sieving of flour for cakes.

Refining: Freeing any material from impurities, as in refining cane-sugar.

Skimming: Removing a floating layer by passing a utensil under it (ladle) as in skimming cream from milk.

Rendering: Separating fat from connective tissues by heat as in rendering lard (dripping).

Filtration: Separating solids from liquids, through fine meshed materials, as in filtering fruit juices for jelly through a cloth bag, or fine wire mesh strainer.

Flavouring: A bundle of herbs and vegetables to impart flavour to stock am sauces is Bouquet Garni.

Folding: Method applies combining different ingredients together so that no reduction in lightness occurs.

Flotation: Separating on the basis of difference in

specific gravity as in the elimination of the over mature peas in a batch by use of brine of appropriate strength.

Evaporation or Reduction: Removal of water, commonly accelerated by heating without lid.

Homogenization: Sub-dividing large drops into smaller ones by forcing them through a small aperture under great pressure as in homogenizing the fat in cream.

Food by Mixing: Food preparation often involves the combining and mixing of different food or food materials. Important effects of the methods of combining food or ingredients are those related to palatability. Texture and flavour are often controlled to an important degree by the skill and method employed in combining component materials.

Emulsification: Dispersing one liquid in another in which it is insoluble. If the dispersion is to be temporary, a stabilizer, which coats the droplets of the dispersed phase, must be incorporated, e.g. in mayonnaise.

Used Methods

Beating: Mixing materials briskly, lifting and dropping them with an appropriate tool. Sometimes used synonymously with whipping as defined below. This is done to thin mixture of liquids. The aim is to mix well and incorporate air.

Blending: Mixing two or more ingredients thoroughly, e.g. blending milk into white roux for bechamel sauce.

Cutting: Usually the incorporation of fat in flour and other sifted dry ingredients with a knife, a method which produces relatively coarse division of the fat and does not result in blending as in cutting the fat into a pastry mixture.

Creaming: Softening fat by friction with a spoon, usually

followed by gradual incorporation of sugar as in cake-making.

Folding: Mixing materials with palate knife or wooden spoon, by a careful lifting and dropping motion as in folding whipped egg whites into a cake mixture. Palate knife is to life.

Kneading: Manipulating by alternating pressure with folding and stretching as in kneading bread dough. A method of combination to the extent that it combines water and flour proteins to make gluten.

Marinating: Coating the surface of food materials, a marinade, which is usually a mixture of oil and acid as in marinating the components of a vegetable salad with French dressing.

Sealing: Sauteing or pre-cooking roast, to develop colour and flavour.

Stirring: Mixing materials with an appropriate tool, such as a spoon by a circular motion, as in stirring white sauce, while cooking.

Whipping: Rapid beating with a wire egg beater or mechanical beater usually to incorporate air, as in whipping egg white.

Whisking: Whisking is done when a mixture needs a lot of air, items need to be mixed together so that they do not separate.

Texture

The texture, as related to food is not an entity in itself, but rather the accumulated effect of several characteristics or qualities to create an individuality such as:

(1) Appearance

(2) Feel to touch
(3) Softness
(4) Mouth feel

Texture is the term used to describe the characteristics of a finished food product. Variety includes some hard and soft food in a meal, so that the amount of chewing required is varied. This is one of the points that is considered while planning a menu. The menu includes dishes that have different textures as soft, crisp, hard, smooth, etc. All the above mentioned factors contribute to the concept of texture.

Appearances: It is the first factor in the appreciation of food. The size, shape and distribution of cells (holes) are of prime importance. Ideally the holes should be comparatively small, slightly oval or elongated and evenly distributed. Large cells (holes) produce coarseness to the eye, as in cakes, bread, etc.

Softness: It is the characteristic of texture and the product should possess the resiliency (springiness) to gradually return to its normal shape--idlis, khaman dhokla, bread, etc.

Feel to Touch: The perceptiveness to touch should be exact as it is desired to be-- a sponge cake should be light and spongy.

Mouth Feel: Texture involves feels or bite tenderness--dry, soft, wet, hard, firmness, crumbliness and short.

Observation, experience, preparing foods, by adding ingredients and incorporating and applying the right principles (combining, mixing, fractionation, etc.) has a definite bearing in the result.

Variety of Textures

Firm and Close: Raising agents added raise the food, the volume is increased, the holes are small and many.

The products are crisp and not spongy, e.g. biscuits, tartlets, etc.

Short and Crumbly: The product is short and it just melts in the mouth, when eaten. The right quantity of fat added gets this effect. More fat is added, as it prevents the mixture from becoming hard and short. It is very similar to firm and close texture, e.g. shortbread, nankhatai, biscuits, short crust pastry, etc.

Light and Even: There are plenty of holes of more or less the same size. It is less short than pastry and less spongy than a sponge cake. The product is firm to touch, e.g. Madeira,-queen cakes, etc.

Spongy: The holes (air cells) are small and evenly distributed, as air has been included. It is soft and elastic to touch as in idlis, khaman dhokla, Swiss rolls, etc.

Flaky: The products have thin crisp flakes and they are formed by air pockets. The crispness is due to the method of rubbing fat with the flour. In order to get a good flaky texture, the right amount of ingredients, proper mixing and correct temperature is essential, e.g. vol-au-vent, patties, bouchees, paratha, tikona mathis, etc.

Smooth: When a dry ingredient is added to a liquid and the blending results in a smooth texture, e.g. sauces, batters, gravies, phirnees.

Incorrect textures spoil a dish and should be avoided. Some of the incorrect texture are:

Coarse and Open Texture: This may be due to insufficient creaming of the fat and sugar, or careless mixing of flour, when added. Too much of raising agents also result in a coarse and open texture.

Hard Texture: The air enclosed has been driven off, may be due to the addition of more liquid than required, or has not been mixed properly. Low temperature of the oven also spoils the texture.

Soggy Texture: This occurs where there is the presence of too much moisture, as it combines the ingredients and the starch gets overcooked and tends to stick, e.g. while preparing rice.

Lumpy Texture: This happens when the liquid and solid is not mixed properly at the same temperature or even if they are at the same temperature. Once the lumps are cooked, they are difficult to remove. This texture may occur in sauces, gnocchi, suji halwas, etc.

2

Method of Food Preparation

> **Every now and then go away, have a little relaxation, for when you come back to your work your judgment will be surer since to remain constantly at work will cause you to lose power of judgment. . . . Go some distance away because then the work appears smaller, and more of it can be taken in at a glance, and a lack of harmony or proportion is more readily seen.**
>
> *--Leonardo Da Vinci*

Heat energy is most basic requirement for preparing food. It can be transferred by three methods--conduction, Convection and radiation. Two or three of these methods may be combined. For example, in baking a cake, converted heat cooks the mixture, then the cake comes in contact with the cooking utensil. Heat is also transferred by conduction. The browning of the cake is by radiation.

Primary Methods

Cooking helps preserve colour, get a variety of texture, enhance flavours and make the nutrients easily available, though a few of the nutrients, which are water soluble are lost or oxidised.

Various Methods of Cooking

1. Boiling

Boiling is cooking by immersing the food in a pan of liquid, which must be kept boiling all the time, i.e. quite a number of bubbles should be seen on the surface. Boiling is restricted for meat and poultry for the first few minutes in order to seal the pores, to retain natural juices and then gentle boiling must take place which is known as simmering at 82°-99°C (180°-210°F). Boiling temperature is 100°C (212°F). Vegetables are boiled at 100°C (212°F). Blanching is used when food needs to be precooked or partly cooked.

Methods

(a) The food items should be completely immersed throughout the process; if excess evaporation takes place the liquid must be replenished.

(b) The flavour of meat and poultry is enhanced by the addition of herbs and vegetables to the cooking liquor.

(c) The liquid in which the food is boiled is known as pot liquor, as it contains some nourishment and flavour. It should not be thrown away but used for gravies or sauces. Use correct ratio of liquid to food.

(d) Scum that arises during the boiling must be removed or it will discolour and spoil the taste.

(e) Fast boiling should be done to green vegetables as it preserves the green colour and prevents excessive loss of vitamins and mineral salts. The pan should be uncovered while boiling.

(f) Root vegetables, with the exception of new potatoes, must be placed in cold water to boil. This helps to improve their flavour, as the acrid taste of root vegetable are leeched out.

(g) Always poach fish and never boil, or the fish will break up. Fish can be wrapped in a cloth and then poached.

(h) Stocks, sauces and soups should be simmered gently. The stock will get cloudy as the scum in fast boiling may get mixed in the pot liquor.

2. *Blanching*

Place food item in cold water and bring to boil, remove. It seals the juices and flavours.

3. *Poaching*

Poaching is cooking gently in water which is just below boiling point (93°-95°C). The liquid is half as close to boiling point as possible without there being any perceptible movement of the liquid.

Methods

(a) Food items should be completely immersed in the liquid and left in throughout the cooling process.
(b) As soon as the liquid, with the food Item begins to boil, lower the temperature to allow poaching.
(c) Whole fish must be placed into cold liquor to poach. Cuts of fish must be placed into boiling liquor, temperature lowered and allowed to poach.
(d) The liquid for poaching eggs at one side of the pan should be simmering and fresh eggs must be added one at a time; this will help to get the rough shape. Placing them into liquid without any movement will result in flat eggs.
(e) It is important that fresh eggs should be used for poaching.
(f) Adding vinegar and salt helps in quicker coagulation and prevents disintegration.

4. *Steaming*

Steaming is cooking by moist heat, i.e. steam--direct or indirect. Indirect steaming is done when the food Is placed in a closed pan, which is surrounded by plenty of steam from fast boiling water or in a steamer. The food item could be protected with grease-proof paper, or cloth or aluminium foil to prevent water getting into the pudding.

Direct steaming is by placing the article in a perforated container or on a covered plate over a saucepan of water.

Fundamental Principles

(a) Only small suitable food items should be steamed, as it is cooked slowly. Green vegetables are not suitable for steaming as they lose the colour.

(b) Puddings in basins must be covered with grease-proof paper to prevent condensed steam from making the pudding wet.

(c) Root vegetables including potatoes must be placed on perforated tray that will allow condensed steam to circulate, and cook them.

(d) It is a good method for invalids, as the steamed foods are easily digested. Nourishment and flavour are retained in the food and usually the food cannot be easily overcooked.

5. Stewing

It is a time-consuming method of cooking, but the advantage is that the coarser, older and cheaper types of poultry and meat are used as they are unsuitable for grilling and roasting. Cheaper cuts of meat and certain fish dishes (bouillabaisse, matelotte) are prepared by this method, as it renders it tender and palatable. Stewing is gentle simmering in a small quantity of water, stock or sauce, until the cut food items are tender, and both liquid and food are served together. Stewing is done in covered saucepans or casseroles on top of the cooker, or in the oven--but always at a low temperature.

Basic Principles

(a) It is a slow process in the liquid or sauce and the long process of cooking will give a concentrated flavour to both the food and the sauce.

(b) While stewing, the liquid should cover the food items.

(c) Various items suitable for stewing are those which are of a tough nature.

(d) The liquid should not reach too high a temperature, as the slow process of cooking by gentle heat, converts the connective tissues into gelatin so that the meat fibres fall apart easily and become digestible. The proteins are coagulated without being over hardened and the soluble nutrients and flavour pass into the liquid which is served.

6. Braising

This is a combined method of roasting and stewing. In a pan with a tight-fitting lid or casserole, braising is done to prevent evaporation, so that the food retains its own juices together with the article added.

Blanching is used when food needs to be precooked or partly cooked for flavouring, e.g. bacon, ham, vegetables and herbs. The meat and poultry used for braising is mature, it is a long method of cooking and tenderizes the less tender cuts and grades of meat. The meat is first browned in its own fat or small amount of hot fat. This seals the pores and thus the juices are retained. It is then placed on a bed of root vegetables. Stock or demiglaze, should come to two-thirds of the meat. The flavourings and seasonings are also added. The lid is put on and it is allowed to cook gently on the stove or in the oven. When nearly done, the lid is removed and the joint is frequently basted to glaze it. This latter process is always done in the oven, to have an even colouring. The above method is the basic method of braising.

The advanced method of braising is by first of all marinating the lardons in the wine and aromatic flavours and then larding the joints, and placing it on a bed of prepared aromatic vegetables in a braising pan and then adding the marinade along with the cooking liquor for braising.

Basic Rules

(a) Braising pan (with handles on either side and tight-fitting lid) should be in keeping with the size of the food item to be braised. This is to ensure even cooking, to avoid hot spots and cold spots.

(b) Fresh vegetables and herbs, with aroma, should be used as a bed for the joint to be braised.

(c) Food items, i.e. meat, poultry and game, should be evenly coloured while frying before being braised.

(d) Braising should be done slowly so as to enable the interchange of flavour between the item and the liquid.

(e) An appropriate liquid should half cover the food item.

(f) Items of meat, poultry and game suitable for braising are those of a tough nature, rather than those of a tender nature.

(g) Braised vegetables are better if served with a good sauce made separately. The braising liquor is strong and similar in taste to the vegetables.

7. *Roasting*

The term roasting was originally applied to the cooking of large pieces of meat, on a turning spit over an open fire, a practice which would be reasonably called grilling or perhaps barbecuing today. Traditional methods of roasting, practised at present are (1) pot roasting, (2) oven roasting, (3) spit roasting, (4) tandoori method of roasting.

(a) *Pot Roasting:* Cooking is done in a covered pot or pan. Only good quality meats, i.e. small joints and birds are used. The meat, poultry or game bird is tied to retain the shape. It is placed in the thick- bottomed pot and skewers placed at the bottom to prevent the meat from sticking to the pan. Fat should cover the bottom of the pan. The joint should touch the fat. Cover the pot with a well-fitting lid and cook on slow fire. The joint should be put in the pot for flavouring after 20 minutes. When the joint is cooked, remove the joint of excess fat. Add stock, so that the juices mix and get the flavour from the aromatic vegetables, and from the basis of the accompanying sauce or gravy.

(b) *Oven Roasting:* It is cooking in a roasting tray in an oven with the aid of fat and first-class meat; poultry and certain vegetables are used. The raw meat joint is trimmed and tied if necessary to retain the shape. It is put in hot fat and all sides are browned to seal the juices. Salt could be sprinkled later on. Baste the meat often, so as to help the meat colour without drying. The average time for cooking is 15 minutes per Ib (455 gm). The time could be adjusted to the size of the joints. For small joints the heat must be constant, and for large joints at first it should be strong heat (450°F) and then reduced to 300°F. After 15 minutes aromatic vegetables should be added for flavouring during the process of cooking. Pork and veal should be well cooked, whereas beef should be just done or underdone. To ascertain the point of cooking, the finger should be pressed against the meat and if it offers no resistance, then it is cooked. Another test is to pierce the meat with a fine skewer and if no pinkish liquid oozes out, it is cooked. The cooked meat should be removed from the roasting tray and the excess fat poured off. The roasting tray should be placed on the fire and the juices allowed to caramelize and the brown stock added and brought to boil. (The aromates and meat juices give flavour and taste to the stock). Thus the brown gravy called *Jus Roti* is prepared and the actual process is known as deglaze.

(c) *Spit Roasting:* It is the original from of roasting.

The same principle of seasoning, basting and cooking with a fierce heat is done. It is cooking by direct heat, basted with fat. A spit is constantly revolved to ensure even colouring and cooking. First-class quality of small pieces of meat are used and the meat pieces should be succulent Quicy). In the case of small game, fuel should be wood, and it should produce low flame.

Basic Rules

(a) Items of meat, poultry and game for roasting must be tender and of top quality.

(b) Food items must be raised off the bottom of the roasting tray to prevent the frying of the lower part.

(c) The juices from the process of roasting meat, poultry and game must be used to make the accompanying gravy.

(d) To facilitate the slicing of roasts, and other cuts of cooked meat, piece should be allowed to remain at room temperature, after it has been removed from the heat for 15 to 30 minutes. This sets the juices, allowing them to soak back into the muscle and it is easier to slice because the meat becomes firm.

(e) The thicker the meat, the lower should be the cooking temperature.

(f) Meat is considered rare when it reaches the temperature of 60°C (140°F); this is the internal temperature of the cooked spots. For well done meat, the internal temperature is 82°-85°C (180°-185°F).

Peeling

It is similar to pot roasting; the difference is that it is cooked entirely in butter. Usually poultry and game are peeled. The poultry is wrapped with thin slices of pork fat, covered with or without buttered paper, placed in a pot and basted with melted butter while it is cooking, covered with a tight-fitting lid. Cut aromatic vegetables are added for flavouring, and later on moistened with Madeira wine (if not available, then red wine could be used.) After the liquor has absorbed the aroma of the vegetables and the cooked food item is removed, it is then strained, grease removed and reduced a little, seasoned and served as the accompanying sauce.

Special peelings are known as "Encasserole" and "Encocotte". The peeling is done is special earthenware utensils and served in the same.

Encasserole

The food item is peeled, only in butter, without the addition of vegetables. When the cooking is done, the food item is withdrawn for a moment and some excellent brown stock is poured. The piece is put into the earthern utensil and kept hot in a bain-marie and served in the same dish.

Encocotte

The procedure is the same as encasserole, except that the food item is garnished with vegetables such as mushrooms, the bottoms of artichokes, small onions, carrots, turnips, etc. which are turned (shaped) and cooked in butter before being used.

Methods

(a) The size of the pot with tight-fitting lid should be in keeping with the item to be peeled.

(b) Sufficient butter should be used to allow for adequate basting and no liquid must be used during the process.

(c) The lid should be removed when the items are two-thirds cooked to colour the food item.

(d) The vegetables, juices and sediments must be used to make the accompanying sauce.

8. Grilling

Cooking by the grilling method began years ago, when meat was first placed near or over an open fire. It has steadily advanced through the years to the most modern gas, the electric grills and infrared models. The food is placed on grill bars, and subjected to the action of radiation which could be from below or above. The source of heat may be from charcoal, coke, gas or electricity. In an authentic grill the heat is located below the grill bars, and if the source of heat is above the grill bars, the grilling equipment is called salamander. The grill bars should be cleaned and oiled, or the bars will rust and food will stick.

Usually small items are grilled, the meat is cut, trimmed and slightly flattened with a bat. Then the meat should be seasoned with salt and pepper and brushed on both sides with oil. The meat is then grilled on the grill bars, turned when half cooked, brushing oil occasionally and cooked to required degree. The fierce heat seals the cut ends of the meat and browns the surface, so that all the juices are retained in the meat. It could be termed as cooking along with concentration of flavour.

French Grilling Terms

Au bleu	- -	rare, very underdone
Ṣaignant	- -	underdone
A point	- -	just done (medium)
Biencuit	--	well cooked *Basic Rules*

(a) Items for grilling must be small in size, flattened and of top quality.

(b) The items should be oiled and seasoned. Fish must be floured, oiled, and seasoned.

(c) Heat should be intense to enable sealing at once and colouring; a brown crust should form, which seals the juices of the food.

(d) Grill bars should be cleaned and oiled.

Broiling is cooking by direct heat and is used synonymously with grilling, It could be done by source of heat above or below. It is dry cooking, and can be done on a grid (iron bars) or pan broiled, where the food is cooked uncovered. The pan or grill is oiled slightly to prevent sticking, excess fat removed.

Stirred eggs (oeuf sur le plat) is an example of broiled eggs. The egg is slipped on to an egg dish, cooked on top of the stove, till white coagulates, then it is placed in a broiler, heat from above. Tenderest and most expensive cuts of beef and fish are most often broiled. Grid markings on steaks add attractiveness. Whole spices broiled on a hot tawa bring out the flavours (pan broiled).

Methods

(1) Tawa or pan or grill may be oiled slightly to prevent sticking.

(2) Dry cooking should take place.

9. Baking

Baking is cooking of food by the action of dry heat in an oven. The degree of dryness of heat may be modified by the amount of steam produced from the items baked. Bread rolls, cakes, pastries, puddings, potatoes, vegetables, farinaceous dishes are baked.

During baking, the mixture or moisture in the product expands by air, steam or carbon dioxide.

Methods

(1) Items should be placed in the right position in the oven, the cool spots should be avoided.

(2) The oven should have the correct temperature according to the baked items, e.g. rich cake should have a lower temperature than pastries, which should have a high temperature. Baked yeast goods should have a hot oven 220°C (425°F).

10. Frying

Food is fried when it is placed or immersed in oil or fat at a sufficiently high temperature; used to brown the surface atleast and partially or completely cook the interior of the food. Frying adds flavour to otherwise bland and tender foods; it also creates a crisp crunchy texture.

There are two types of frying.

(a) *Deep Frying:* It is a quick method of frying and a popular method for Indian cuisine. The holding temperature range for deep-fryer is 90°C-110°C. As the food is immersed in hot fat, it is essential that the fat is of a good quality having a high smoking point. The frying pan should have a sufficient depth. The size and shape of the food to be fried should be uniform. Fried food have a good appearance. Some of the fried food are: rissoles, cromesquies, chicken and mutton cutlets, fish cakes, potatoes, vegetables, fritters, beignets, croquettes, pakodas, macchi pakodas, vadas, moong dal ladoos, samosas, etc.

Methods

(a) The fat must be hot enough to seal the outside of the items being fried. This prevents the food from absorbing any fat and to prevent any loss of flavour. The exceptions to this rule are for the frying of Dauphine, Chamonix potatoes for which the temperature is approximately 145°C to allow the slow expansion in size while cooking. Frying temperature varies from 160° to 195°C.

(b) Prevents the escape of juices; all items, except potatoes must be seasoned and coated with either batter, flour, egg or egg and breadcrumbs before being fried.

(c) Too many items should not be put at a time as it will reduce the temperature to the point where the food will absorb fat, lose its juices, become soggy and heavy or break up.

(d) The fried items should have a uniform golden colour.
(e) Drain well on paper and serve at once on doilys.
(f) The fat used for frying should be strained and stored in a cool place.
(g) If fat, that has been used for cooking, is used again, it should be replenished with fresh fat for reuse in the ratio of 1:2.
(h) Wet food items, e.g. potatoes should be well dried before being fried.
(i) If possible, a frying basket should be used to give uniform colouring and heating to the items fried.

(b) *Shallow Frying:* Tender, good quality items, like fish, meat and poultry are generally used for frying, as this method completes the cooking of the dish. The presentation side i should always be fried first and then turned, both sides are browned evenly. For bacon and sausages as they have sufficient fat, do not add additional fat. The food items, that are shallow fried are fish cakes, medaillons, eggs, parathas, dosas, cheelas, alloo ki tikkies, etc. Sauteing comes under shallow frying--the word sauteing comes from the French word *sautoir* which means "to jump". Food items, e.g. thin meat pieces, fish, game and vegetables are cut into very small thin pieces. They are placed in a heavy-bottomed pan with a little hot fat and the pan with the items is shaken and flipped occasionally over the fire.

Procedure

(a) Items should be seasoned before being shallow fried.
(b) Food items must be placed to cook in hot fat, so as to seal the outside.
(c) Drain well the food items after cooking, or the food will be greasy, e.g. bacon rashers.
(d) Before sauteing, food having a high water content

should be salted and allowed to drain and then dried on paper towels. (Salt draws out the moisture).

(e) The saute pan and frying pan should have a thick bottom and straight sides.

Solar Cooking

It is similar to what we use in the kitchen to cook food. Solar cooker works only on *solar energy,* it emits no smoke, no soot spoils the cooking utensils. It keeps the surroundings clean. Above all, it conserves the precious energy resources of the country and saves us money.

Principles: Solar cookers can be broadly divided into two categories:

(1) Cookers utilising solar heat with little or no concentration of rays. It may be termed as a "hot box".

(2) Cookers which deflect solar energy from a large surface to a smaller area to produce high temperatures.

The first type of solar cooker commonly known as "hot box" consists of a well-insulated box, the inside of which is painted dull black and is covered by one or more transparent covers. The purpose of the transparent covers is to trap heat inside the solar cooker. These cookers allow the radiation from the sun to come inside but do not allow the heat from the black absorbing plate to go out of the box. It is because of this that the temperature of the blackened plate inside the box increases and can heat up the space inside to temperature up to 140°C which is adequate for cooking. The second type of solar cooker uses a lens or a reflector suitably designed to concentrate the solar radiation over a small area. This cooker is able to provide high temperature on its absorbing surface when suitably designed.

The cooking time is about 2-2% hours for cooking depending upon the kind of food and the season. The time taken for cooking is less in summer than in winter.

Merits: The solar cooker has a number of advantages over the traditional cooking devices.

1. It cooks up to four items at a time.
2. It preserves the nutrition value of the food because the cooking is done at low temperature.
3. It does not require constant attention.
4. It is safe and does not burn.
5. Saves fuel.
6. Keeps food hot for a long time.

Disadvantages

1. It needs the sun and will not give results on a cloudy day.
2. Only simmering method is adopted.

Solar Cooker

1. Outer box of a solar cooker may be made of wood, iron sheet or filter fibre reinforced plastic having suitable dimensions.
2. The inner box can be made from galvanized iron, or mild steel or aluminium sheet. All the four sides and the bottom of the inner box which are exposed to the sun are coated with blackboard paint which absorbs solar radiation.
3. Thermal Insulation: The space between the outer box and the inner box is fitted with insulating material such as glass wool, thermocole, etc. This prevents heat losses from the cookers.
4. The double glass cover: Generally a double glass cover is provided for a solar cooker. These covers have length and breadth slightly bigger than the

inner box and can be fixed in a wooden frame maintaining a small spacing between the two glasses. This space contains air which acts as an insulator and prevents heat escaping from inside. The frame is attached by means of hinges. A rubber strip affixed all around prevents any heat leakage.

Mirror: It is used in a solar cooker to increase the radiation input on the absorbing surface. Sunlight rays on the mirror gets reflected from it and enter the box after passing through the glass covers. This radiation is in addition to the radiation entering the box directly and helps to quicken the cooking process by raising the inside temperature of the cooker.

Inferred Cooking: Dry heat is excellent for cooking meat, fish and other items which are fairly tender to begin with. The food is placed a few inches above or below the heat source. Some of the heat which is given off by the heat source is carried to the food by means of convection, the rest by means of radiation. The radiation is in the form of inferred rays. Inferred rays are a form of radiation which are just a little longer than those waves which are seen as red and shorter than radio waves and microwaves.

Inferred wavelengths are small and range in length from about 0.07 to 1000 millionths of a metre. Inferred waves are energy waves just a little longer than light waves. Wavelengths from 1.4 to 5 millionths of a metre are the most effective for cooking food, because they pass through the vapours and act on the food and the moisture contained in the food. Shorter wavelengths are not well absorbed by the water in the food.

Catering equipment which relies on inferred waves for energy transfer should produce inferred waves from 1.4 to 5 millionths of a metre.

As temperature increases in a heat source, the energy radiated shifts to shorter-wavelengths. As the temperature is doubled, the radiated energy shifts maximum to half the wavelengths. A temperature of 649°C (1200°F), for example produces a maximum energy wavelength of 3 millionth of a metre effective in inferred cooking.

However, the surface of food cannot stand temperatures much above 260°C (500°F) without burning (unless they are very moist). With high temperatures, food must be placed some distance from the heat source. High temperatures are desirable to produce inferred waves of a maximum cooking effectiveness but are so hot that if food is placed near this heat source, it will burn.

Infrared heating is fast because the energy is transmitted from the heat source to the food almost instantaneously, not having to be carried by means of hot air to the food, as in the case of convection heating.

The infrared rays penetrate the blanket of vapour surrounding the food and enter the food a few millimetres which also speeds up the cooking process.

Vapour Blanket Food: Some of the new grills produce a great amount of inferred energy and hence are much faster than the old style grill which produced less inferred energy.

The closer the food is placed to the source of radiation, the faster it would cook. Radiant heat energy is transmitted directly to the food passing through the vapour blanket. The amount of energy which reaches food from an inferred source varies proportionately to the distance the food is away from the source.

A steak placed 5 cm away from a heat source will cook four times as fast as the same steak 10 cm from the source. Infrared rays travel in straight lines in all directions from the emission source. Aluminium foil should not be placed on the food as the rays penetration will be less.

Infrared lamps are also used to keep food hot and to give them a rich red colour on buffer tables and cafeteria lines. Inferred lamps transmit heat instantaneously to food exposed to them and make the food appear more red in colour.

The amount of heat absorbed by food from an inferred source depends upon the reflectivity of the surface of the food; the darker the colour of the surface of the food, the greater the rate of absorption.

Infrared lamps and inferred grills are designed so that most of the energy emitted is inferred.

One infrared grill cooks a 340 gms steak in about 6 minutes-- less than half the time required for grilling in a conventional grill. Infra-red energy is produced by electric cooking as well as by gas cooking.

The ultra-ray is gas fired and is a recent invention and produces 100% inferred radiation. It consists of three layers of Incaloy metal screens sandwiched together. Air and gas readily pass through the screen, produce a blanket of flame about 1 cm in thickness.

Temperatures range around 621°C (1150°F), reaching maximum capacity within six seconds.

With such a grill, time required for grilling is determined by the distance from the energy source and its temperature.

Haute Cuisine has some special methods of cooking which is done when required.

Sous Cloche (Under bell cooking): The name is derived from a bell container, made of glass or silver. It has two parts, a base and a lid. The dish is cooked in the cloche and it is served in the same container. The bell-shaped lid is opened in front of the customer.

Champignons Sous Cloche: The ingredients are: Mushrooms, white wine and butter. Heat the butter in the base of the container, add sauted mushrooms, wine, seasoning. Place the lid on top, cook for 2 minutes. The container is placed in front of the customer and the waiter removes the bell-shaped lid (all the essence and goodness is retained).

En Papillate: This is a traditional method of cookery, where the article of food is placed inside oiled grease-proof paper cut according to the shape of the food.

It is generally applied to items that need less cooking time, e.g. tender cuts (prime quality of meat), fish or vegetables that need less cooking time.

Meat when cooked by this process is set rapidly by shallow frying, without cooking fully. This is done to seal natural juices. Only tender meat is used for this method such as minute steak, kidneys; liver, etc. Generally a garnish is placed on the oiled greaseproof paper which is wrapped round the article of food and the edges sealed by folding them over. The bag is then placed on the service dish and cooked in the oven. The steam that rises within the paper container makes the top surface of the paper rise, e.g. Cote de Vean en Papillote.

Prepare as follows: Fold the grease proof paper as illustrated, well covered with oil. Place on the paper, a thin slice of ham (pear-shaped) the same size as of a cutlet. Spread duxelle (finely chopped mushrooms, onions, carrots) in the ham slice. Then lightly saute the prepared seasoned veal cutlets for 2 minutes on each side. Place the cutlet on the duxelle that has been put on the ham slices, then place a slice of ham on top.

Take grease-proof paper, fold into two, cut heart shape with scissors; it should be larger than the food item. Open the cut paper, on one side and place the food item as explained above. Seal the edges of the paper from point A in the diagram and fold at 20 mm intervals, flattening each fold, before the next fold (see diagram).

Sous La Cendre: Meat slice is larded, seasoned and marinated in white vinegar. Wrap in a paste covering the meat. Place in the oven after making ½ inch hole. (This is done to let off the steam.) When half baked, put demi-glaze in the hole, finish baking. Serve hot with Madeira sauce.

Forced-Air Connection Oven. The forced-air connection oven is best suited for use in a kitchen either as a cooking agent or a reheater prior to presentation of the food on a counter service. The advantage is that an even temperature is maintained throughout the cooking process but a high degree is reached in a relatively short time, i.e. 390°F in 15 minutes. This is achieved by a centrifugal fan drawing air across the heating elements and distributing it evenly. The movement of air strips away the dead air which normally surrounds a product when first put into the oven and promotes faster heating up.

At the same time because of the air movement taking place, the temperature throughout the whole oven remains constant, whereas in conventional static atmosphere units there is a tendency for temperature gradients to be developed. The oven can roast, grill, bake, thaw and cook quickly and efficiently. The advantages of convection oven is two fold, it is not only speedier, it reduces the weight loss incurred in the longer cooking cycle with conventional ovens.

Forced air convection ovens are also eminently suitable as end cooking units for prepared frozen meals coming individually or in bulk. The food can be placed in the oven straight from the zero cold room and end-cooked within 20 to 30 minutes. The smaller type of forced air convection oven is suitable for such installations as snack-bars, bakeries and in public houses where there is a demand for hot dishes.

This system could also be applied in the kitchen having satellite service points, snack-bars, bakeries where there is demand for hot dishes.

Microwave: Although a form of energy, it is not heat. The principle of microwave involves a, type of electronic heating and is concerned with the production of high frequency waves from an oscillator tube called a magnetron. The electromagnetic waves have energy and this energy is transported to the desired load for heating. The energy created penetrates articles of food to a depth of 3" and causes the molecules that make up the food to rub together and create friction, supplying the heat which cooks or reheats the food.

Microwaves do not heat the oven cavity. Microwaves are reflected by metal items and therefore are unsuitable for use with this equipment.

The advantage lies in the speed of the operation, although browning is not possible. It is excellent as a reheater. Microwave ovens allow direct customer participation with the use of time tokens.

Electronic Cookers: The hotter a substance becomes, the more rapidly its molecules or atoms begin to vibrate. The common methods of applying heat by conduction, convection or radiation, all involve supplying heat energy to excite.

Electricity can be changed into microwaves using a special valve called a magnetron. The microwaves pass through air or glass, china, plastics and paper without heating them, but when they strike solid and liquid foods, they are absorbed. Their energy is transferred to the food molecules which vibrate rapidly. The resultant heat cooks the food. Microwaves are absorbed by most types of food in varying degrees dependent on the cellular structure of the food.

Affecting Factors

(a) *Density:* The denser the food, the more difficult it is to cook. A joint of red meat will not be easily and uniformly penetrated by microwaves whereas liquids, fish, white meats, etc. offer far less resistance.

(b) *Quantity:* The heating is proportional to the quantity of food placed in the oven. For instance if one portion takes 30 seconds to heat, two portions will take 45 seconds, whereas three portions will take 60 seconds. For each additional portion, half the time taken by the first portion is added.

(c) *Shape:* The thicker the product, the longer it will take to cook. Shape as far as possible should have uniform thickness, e.g. truss chicken.

(d) *Containers:* They should be of glass, china, plastic or paper to allow transmission of microwaves.

1. Microwave heats the food by agitating the food molecules.
2. It is a rapid method for small quantities and small portions of uncooked food.
3. There are no food odours and, less steam.
4. Excellent method for reheating, frozen cooked food, canned food for reheating assembled food on a plate or pre-cooked snack meals.
5. Rapid service and needs less staff.

One disadvantage is that food does not brown; browning elements can however be fitted, if required.

Rechauffe: Rechauffe means to reheat. Frequently many cooks come across a few leftovers in their kitchens. In the interest of economy, a sound knowledge of the left-over foods is necessary. These left-overs should be used in some form or the other. Many tasty and interesting dishes can be prepared, but care should be taken that the food is thoroughly and carefully reheated. It is also important that only sound food is used. "If there is any doubt, throw it out." If meat is over-cooked, it will toughen the fibres and harden the proteins thus making food indigestible and unpalatable. Meat and fish lose some of their nourishing components in cooking and this loss could be supplemented by other food. In addition to these points, the importance of suitable seasonings and flavourings and suitable accompaniments adds and supplements any deficiency in food value, and also determines the success and value of a rechauffe dish. Care must be taken that frequent repetition does not occur. Always the use of minced meat or mashed potatoes will make it monotonous. In producing a dish with left-overs, it should not be obvious that the main ingredient of such a dish is not fresh.

Rechauffe dishes can be made attractive, if prepared and served in small individual dishes or patty cases and fish in small scallops or coquille shells. The dishes must be decorated so as to look attractive and pleasing to the eye, It should be combined again to enhance the flavour with a sharp sauce as the natural flavour is lost to some extent. Food may be warmed in a moderate hot oven carefully covered. The vessels containing food, if placed in a tray of water in the oven, prevents over-drying. Steaming is one of the best methods of reheating as the heat does not dry or over-cook the food.

Same Guidelines

1. To bring down the food cost, utilize the left-overs, such as vegetables, eggs, rice, cheese, sauce or gravies, etc.

2. The left-over vegetables could be mixed with salad dressings, used as salads and served cold.
3. Sauces could be made sharp by adding pepper, mustard, vinegar or tomato puree, etc.
4. Panada or thick binding sauces will help to bind mashed or chopped ingredients for croquettes. Mashed potatoes could be used without sauce, fish cakes, aloo ki tikkis, aloo ka parathas, etc.
5. Cooked left-over rice could be made into croquettes, lemon or curd rice.
6. Curries could be improved by adding fresh ingredients such as beaten curds or ground cashewnuts or coconut paste. Onions, mushrooms, curds, curry paste improve the flavour and taste when added.
7. It is important that the excess fat, gristle is removed or it will make the rechauffe dish greasy.
8. Trimmings and bones could be used for brown stock.
9. White sauce is appropriate for fish and poultry, and brown sauce for the meat.
10. Marinating of meat or fish improves the flavour.
11. To improve the taste for fish, white meat, lemon juice, parsley, nutmeg could be added.
12. The flavourings and seasonings, when mixed with finely minced meat, blend better and heat penetrates quickly to all parts and this reduces the reheating time. Fish can be flaked and blended with spices.
13. The addition of moisture in the form of gravy or sauce is necessary for the reheated meat dish, whether it is incorporated inside or served separately as it gives moisture to the dish and improves the taste.
14. If raw ingredients are to be added to the rechauffe dishes, they must be cooked first, as the short time necessary for reheating does not allow time for raw

food to be thoroughly cooked, e.g. onions in croquette mixture, potatoes in fish cakes, rice in cutlet mixture.

15. Protection of the food from direct heat of an oven or from hot fat can be done by giving protective coating by enclosing with batter, egg and breadcrumbs or covering with mashed potatoes as in shepherd's pie.
16. Left-over stale bread can be used to prepare breadcrumbs and puddings, cheese toast.
17. Rechauffe dishes should be dished neatly, decorated and garnished.
18. It is important that the dishes should be piping hot. *Reheating Dishes*

Meat: Any meat or mixture of meats--white or red--poultry, game or rabbit can be used with suitable flavourings and seasonings for croquettes, cutlets, rissoles, kromesky, salads/ cottage pie, mirotan of mutton, shami kababs, keema pulao, etc.

Croquettes: Make the panada sauce, chop meat finely, otherwise the mixture cannot be smooth, add seasonings and flavourings, add to the panada and mix well. Turn the mixture to get cold and firm. Make rolls of even size of the mixture. Coat with egg and breadcrumbs. Press the coating well and coat a second time. This improves their appearance. Fry in hot deep fat, drain well and serve with fried parsley.

For Cutlets: Form minced meat into cutlet shapes, season, egg and breadcrumb twice, put a piece of macaroni in the end of each, and deep fry. Serve with fried parsley.

For Rissoles: Roll out short crust pastry very thin and cut into rounds about 4 to 4½" in diameter. Place a little of the prepared mixture on each, roll and seal edges, coat with egg and crumbs, fry in deep fat.

Kromeskys: These make a dainty small entree dish.

Just the scraps and trimmings of white meat, ham or tongue can be used. Prepare coating batter, mince meat and mushrooms finely, add seasoning, grated lemon rind and mix with sufficient panada. Mix well, form into small rolls and wrap each in a small slice of bacon or caul. Dip the rolls in batter. Fry in moderately hot fat till crisp and golden brown in colour. Garnish with lemon and parsley.

Chicken: Left-over chicken cut into joints can be reheated carefully in sauce and if the skin and bones are removed, the chicken can be used with bechamel sauce in patty cases or croquettes. Cold roast chicken could be used to prepare chicken sandwiches and canapes, chicken chat, etc.

Fish: Cooked fish is more insipid than meat and requires to be seasoned well and flavoured. Lemon juice, parsley, anchovy essence are the flavourings used.

Cold fish may be used in may interesting dishes, e.g. fish mornay, curried fish, fish cakes, fish cutlets or croquettes, fish pie or fish salads.

Fish Cakes: Flake the cooked fish, add boiled potatoes (two-third fish and one-third potatoes), chopped parsley, lemon juice and seasoning. Add a little melted butter, egg to bind and cool the mixture. Make round cakes, egg and breadcrumb twice and deep fry. Garnish with lemon and fried parsley.

Fish Croquettes or Cutlets: Flake fish well, season, mix with panada to bind and then roll, egg and breadcrumbs and deep fry.

Fish Scallops: Butter some china scallop shells or small deep round bowls, place the flaked fish that is seasoned and mixed with lemon juice. Cover with white sauce or sprinkle breadcrumbs and melted butter on top. Pipe the mashed potato all round the fish.

Reheat in a quick oven or gratinate, till the surface is nicely browned. *Fish Pakodas:* Dip fish in besan (gram) batter and deep fry.

Vegetables: Cold cooked vegetables such as peas, beans, cauliflower, potatoes, mixed with vinaigrette or mayonnaise are used as a salad. Cold boiled potatoes can be used for potato salads, tikkies or for saute potatoes. Mashed potatoes are used for fish or potato cakes, or for making borders for salads, or covering meat cutlets. Rice, cooked spaghetti or macaroni can be used for hors d'oeuvres with chopped onions, tomatoes, cooked meat, vegetables mixed with vinaigrette or mayonnaise.

Bread: Trimmings of crust, etc. should be kept until dry, lightly browned in the oven and made into breadcrumbs. Stale bread can be used for bread pudding, shahi tukra, bread halwa, pakodas, etc.

To control the cost of food, the left-overs in the kitchen are--rice, mutton, gravy, vegetables, etc.

Remember econmy and wise use of leftovers can fetch in profit. A skilled chef keeps all these points in mind.

3

Fundamentals of Menu Planning

> Research is an organized method of trying to find out what you are going to do after you cannot do what you are doing now. It may also be said to be the method of keeping a customer reasonably dissatisfied with what he has. That means constant improvement and change so that the customer will be stimulated to desire the new product enough to buy it to replace the one he has.
>
> —*Charles F Kettering*

In conventional sense, menu is a list of dishes served at a meal or that can be ordered in a restaurant.

Menu planning is the technique for composing a series of dishes for a meal. Composing a good menu is an art and it needs careful selection of dishes for the different courses, so that each dish harmonises with the other.

The planning of meals in commercial catering establishments is based more on economic considerations and reputation than on the desire to provide nutritionally "balanced diets". The dishes produced are intended to please the eye and the palate. The planning of menus for school feeding, industrial canteens, hostels, etc. has a different aspect, and nutritious, and well balanced foods are compiled.

For achieving success there are some rules in the gastronomic laws which must be observed whereas a badly compiled menu will spoil the complete meal. Menu should provide nutritious food, tempt the appetite and satisfy the guests.

Popular Courses

There is no exaggeration in admitting that it is essential to have a knowledge about the sequence of courses in Western menus, because for Indian menus all the dishes are served at one time in a thali. The modern trend is to give about 4 to 6 courses and a list of various courses has been given. This is to enable one to choose from the sequence.

Sequence of Courses

1. Hors d'oeuvres
2. Soups--thick or clear
3. Fish
4. Entree
5. Releves
6. Sorbet
7. Roasts/Grills
8. Vegetables/Salads
9. Sweet
10. Savoury
11. Dessert (fresh fruit, nuts), Coffee (Not a course)

Many types of menus are there which include Table d'hote. Carte de jour, A la carte, special party menu for banquets, buffets, parties--formal and informal.

(a) Table D'hote (Table of the Host)

A fixed menu planned for a complete meal at a fixed price. It is a meal of three or four courses with a limited choice of dishes, and it is cooked in advance. It is changed daily or may be used in rotation. It is set in the classical sequence starting with hors d'oeuvres or soups and so on.

Left-over food items could be used for this type of menu as it helps to economise.

(b) Carte De (Jour Daily Card)

These are sometimes an insert in an a la Carte menu, which comprises a few special dishes of the day to vary a la Carte menu, for regular customers. Some establishments have a set pattern for their Carte de jour, e.g.

Roast Chicken	Monday
Fish and Chips	Tuesday
Curry and Rice	Wednesday

A La Carte (by the Card)

A la carte means the presentation of a menu of a long sequence of dishes and courses. Each dish is individually priced. Dishes are cooked to order and the portions are large than a table d'hote menu. The customer has a wide choice and should be prepared to wait for this service.

Many dishes and sauces are prepared in advance except for the final finishing touches and cooking. This type of menu is normally permanent, until either prices or change of management makes it out of date.

Banquet

Usually the menu is composed for a number of people and it is necessary to know the reason for the banquet. It is usually a formal affair and dishes should be compiled to suit the occasion. The dishes chosen should be such that will not require last minute dressing. The menu is elaborate and of a high-class quality. It is fixed menu, with no choice. It is not usual to include farinaceous dishes or eggs and stews. The service as a rule is faster and takes the following form:

First Course -- Fruit cocktails, smoked salmon, hors d'oeuvres or soups.

Second Course -- Fish--Fish Bonne Femme, Fish Duglere, etc. but without the complicated garnishes.

Third Course -- Entree (excluding stews and rechauffe dishes), roasts, grills or cold meats with vegetables and potatoes or salads.

Fourth Course -- Sweet dishes with emphasis on the hot variety, or Bombes.

Fifth -- Cheese and biscuits.

Course Coffee -- Always served.

Buffets

Generally, there are two main types of buffets, e.g. the Light Buffet and the Fork Buffet (lunch and supper). This buffet table is attractively displayed. The food served is colourful and attractive, tastefully decorated. A large variety of dishes is offered for everyone to choose.

Light Buffets are give for various reasons to augment a long function, i.e. late in the evening, also could be given for tea dances and supper dances, etc.

The Fork Buffet will also provide a variety of dishes and the food item can be easily eaten with a spoon or fork, while standing. The popular dishes chosen are chicken mousse, ballotine, terrines, salads, etc. A list of dishes that can be offered for buffets are-- savoury finger toasts, cold or hot canapes, chipolata, sausages, cheese, game chips, patties, sandwiches, smoked salmon, eggs, ham, trifles, jellies, fruit salads, ice-cream, pastries, gateaux, etc. Coffee, tea or punch.

The modern trend is buffet lunches or dinners. The elaborate well decorated dishes are displayed. The price is fixed. Western and Indian dishes are displayed together and the guests can take both or either.

The main dishes displayed are--

Salads, soups, roasts, fish fried, vegetables, ragouts, fricassees, potato dishes, and puddings of different variety. The Indian dishes are pulaos, biryanis, kormas, roghanjosh, jhalfareze, bhajees, bhurtas, raitas, papads, pickles, salads, nans, bhaturas, chole, tandoori chicken, seekh kababs, etc.

Cocktail

In this the main ingredients are actually drinks. Very tiny savoury snacks are served such as stuffed olives, gherkins, walnuts, game chips (wafers), salted almonds, chipolata, sausages, bacon rolls, cheese fritters, cheese straws, canapes, bouchees, shami kababs (small), mini seekh kababs, aloo- ki tikkies, chiwdas, etc.

How Menus are Prepared?

Type of Meal: The different types are--Breakfast, luncheon, dinner, supper, tea, etc.

Breakfast: There are three types--Continental, English and Indian.

Continental: It is a light breakfast, comprises brioche, bread, croissants or French loaf, jam, jelly or marmalade and coffee or tea.

English: It is on a larger scale and comprises fruit--fresh of stewed, cereals, bacon and eggs to order, or fish, preserve, breads and beverages.

Indian: Popular breakfast comprises stuffed parathas and curd or potato bhajee and poories, or besan cheelas and curd or dosas and chutney and pickles, and beverages like tea, coffee, lassi or milk.

Luncheon Menus

Menus are usually shorter than for dinner with less courses and simpler dishes, but more choice within each course. The emphasis is placed on stews, roast joints, grilled meats, cold buffets and steamed puddings. Luncheon is a quick affair and dinner is taken more leisurely. A six-course sequence of dishes is given:

Sequence of Dishes

First Course -- *Appetiser*--Fruit or shellfish cocktails, delicacies such as caviar, oysters, etc. smoked fish or hors d'oeuvres.

Second Course -- *Soup*--Thin or thick soups.

Third Course -- *Fish*--Without complicated garnishes. The fish could be steamed, grilled or fried.

Fourth Course - - *Meat / Chicken*--Stews, blanquettes. fricassees, goulash, cutlets, rechauffe dishes, braised or roasted meat. Served with vegetables and potatoes, Salads can also be served.

Vegetables--A choice of vegetable dishes is given for vegetarians such as augratin, cutlets, vegetables bombe, kievs, etc.

Fifth Course -- *Sweet--Hot* puddings, pancakes, pies, flans, milk puddings, ice-cream, fruit salads, bavarois, cold souffles, etc.

Sixth Course -- Dessert--Fresh fruit, nuts, etc.

Coffee

Dinner Menu

Dinner menu comprises highly garnished, classical dishes. Here one has a scope of showing one's skill of compiling menus with a choice of the most esteemed and rarest and exotic of dishes. Large joints could be carved in the restaurant in front of the customers or fiambe dishes could be served. Balance must be perfected throughout the long complicated number of dishes.

Dinner Menu Courses

First Course -- *Appetiser*--Cocktail, caviare, oysters, snails, etc. Smoked salmon, foie gras, hot hors d'oeuvres.

Second Course -- *Soups*--Consomme, cream, soups, bisques, bortsch, etc.

Third Course -- *Fish*--Deep poached salmon, trout with sauce and garnishes. Coquilles St. Jacques (shellfish cooked with sauce served in a shell), grilled lobster, lobster thermidor. Fish meuniere with garnish for sole, turbot, trout, salmon. Fried whitebait. Cold fish always in aspic.

Fourth Course -- *Entrees / Releves*--Tournedos, noisettes, vol au vents, large joints, chicken poele, grilled beef fillet, chicken en casserole, encocolte, served with vegetables and potatoes.

Fifth Course -- *Sorbet*--In a long menu, diners refresh their appetite with a light sherbet, flavoured with champagne or liqueur.

Sixth Course -- *Roast*--Roast game and poultry served with a salad.

Seventh Course -- *Sweet*--Cold or hot--souffles, pancakes, crepesuzettes, melba ice-cream, baked Alaska.

Eighth Course -- *Dessert*--Fresh fruit and nuts.

Coffee

Indian Menu

As a unique feature, there is no difference between lunch and dinner menus. All the courses are served together on a thali. A list of Indian dishes can be as follows to make the planning of menus easier:

Meat	--	Chicken badam, pasanda chicken, tandoori, gosht kaliya, korma, roghanjosh, keema matar, mutton koftas, pork vindaloo, muttons baffat.
Fish dishes	--	Fish Patiala, patiya, masala machhi, machhi jhal, Goan fish curry.
Vegetarian dishes	--	Malai kofta curry, Aloo dum Bhojpuri, paneer jhalfareze, paneer pasanda, paneer matar.
Lentil dishes	--	Rajma shahi, sabut urd dal makhani, khatte chhole, Peshawari chhole, besan kadhi, sambar, amti.
Sukhi sabzi	--	Aloo matar, bhindi masala, karela stuffed, brinjal Patiala, Potato and tomato, beans foogath, Gobi Mussallam.
Raitas	--	Mint, potato, tomato and onions, boondi, bhallas and gunjia.
Rice dishes	--	Tomato and coconut pulao, zer biryani. Kashmiri pulao, keema pulao, puliodarai, lime rice, curd rice, aloo ki tahari, pea pulao, etc.
Indian bread	--	Nans, parathas, roomali roti, tandoori roti, bhaturas, kulchas, Poories, kachories, etc.

Supper: A full buffet is given for supper. All the dishes are very well decorated, giving a good display.

Menus For When?

Menus will vary according to the type of establishment:

(1) Hotels
(2) Restaurants
(3) Hostels
(4) Hospitals
(5) Industrial / Office canteens

The menus for restaurants and hotels will be A la Carte, Buffet. Banquet, or Table d'hote. The dishes will please the eye and palate and will be determined by the customers' appetite and pocket. On the other hand the hostels, hospitals and canteens will have a set pattern and will offer meals to provide nutritionally balanced diets within economy budget. As there is no choice and food is eaten daily, a variety given will minimise the monotony and food preference must be seen.

(a) Customers' Influence

Customers can influence the type of food served because of the following factors: age, sex, occupation. The young boys will need substantial meals, whereas the young girls would prefer dainty and small portioned dishes. A group of farmers, who are used to eating well, will need a good quantity of food, and the emphasis will be on palatability and wholesome food.

FACTORS AFFECTING MENU PLANNING

(1) Capabilities of Catering Staff

The kitchen staffs capabilities have to be seen, whether they will be able to cope with the high-class cookery, whether they are experienced and have the skill and the knowledge. If the staff is not capable, it will be difficult to produce good meals. Equally important is the serving staff. If they are efficient, are well dressed and garnished, dishes could be served in a neat and presentable manner. If the waiters are untrained, the selection of dishes should be done carefully. Varied and sophisticated dishes can only be given if the waiter is highly skilled.

(2) Occasion

Special dishes for certain days or time of the year should be considered. For example, roast turkey at Christmas, special eggs for Easter, halwa for Baisakhi, sweet rice for Basant, sheer kurma for Id, are the seasonal and occasional foods.

(3) Seasonal Availability of Raw Materials

Season is important in the choice of food. Cool, crisp and fresh foods are ideal in summer. For the cold weather, heavier, richer foods high in calorific value are preferred. Although in these days of cold storage, foods are available all the year round, foods in season should be included in menus, as they are fresh, plentiful, and the colour and flavour are good. They are available at a reasonable price. Foods selected should be easy to obtain locally.

(4) Kitchen Equipments

While planning the menu, it is important to see that the kitchen is well equipped as to be able to cope up with the preparation of various dishes. Care should be taken to see that the method of cooking is not repeated; otherwise certain equipments would be overloaded.

(5) Price Menu

Always give value for money. The food cost should not exceed 40% of the selling price at an average, e.g. if the food cost per head is Rs. 2/-, then the suggested selling price would be--

$$\frac{2 \times 100}{40} = \text{Rs. } 5/-$$

$$\frac{\text{Cost of food} \times 100}{\text{Gross profit subtracted}} = \text{Selling price of dish from } 100 = \text{food percentage.}$$

The proposed charge per head is obviously an important factor to consider for selecting food for any menu. If the price of the meal is Rs. 30/- expensive items cannot be considered. If asked to produce meals for a good price, say over Rs. 100/- per head, then good value for the price charged, should be given.

Writing Menus

This is particularly important, when compiling menus and special menus, the following points should be considered.

(a) Repetition of Colour: Wherever it is possible, this should be avoided to make the menu exciting. If the menu consists of celery soup, fricassee of chicken, turnips and creamed potatoes, vanilla ice cream, which are all white, it is bound to look dull.

(b) Repetition of Ingredients: The basic ingredient on the menu should never be repeated as it becomes monotonous. For example, if the soup is of tomatoes, then the tomatoes should not reappear in any other course on the same menu. When using classical garnish in the menus, see that any vegetable, which is included in the garnish, is not repeated as a vegetable.

(c) Repetition of Words: Always avoid the repetition of the same word on the menu, as it seems the planner has a limited knowledge and makes the menu less interesting, e.g.

Cream de tomate

Fried fillet of sole

Grilled fillet steak

Creamed potatoes

Fraisse a' la creme.

Avoid Overbalance of a Menu: If many courses are served, then care must be taken to ensure that they are neither too light nor too heavy.

Texture of the Courses: Some food should be soft whilst other food should require thorough chewing. Avoid menus with the same texture.

Seasoning: Do not over season the food. If strong herbs are used in one dish, never repeat them in another dish in the same meal.

Garnishes: These must be correct. Garnish attractively and simply.

Colour: Sensible use of colour in dishes will give them an "eye appeal". Deep vivid colours should be avoided. Drab looking dishes can be improved by an attractive, colourful garnish.

Food Value: Special attention should be paid to the different requirements of different groups of people. The manual worker requires more substantial food than an office worker. Meals served to children, old people, invalids, adolescents, and expectant mothers should be nutritionally balanced.

Wording of Menus: When compiling menus, the following points must be considered:

(1) Select language which the customer can understand.
(2) If the menu is in French, give an English translation.
(3) Never use a mixture of languages for the courses of a menu.
(4) Spellings of the words should be correct.

Menu Merchandising: Menus are planned and designed. Menu should reflect the policy and thinking of the management and should ensure profits.

The menu should be suitable for the market and the ability of the employees if the operation is to succeed.

The presentation of the menu should be taken care of Menu card should be clean and not be dirtry or frayed as it reflects on the management. The menu should be legible and readable for the customer. The format of the menu should suit the contents and different format should be used for different meals. Menu should be organised, similar items should be written under attractive headings and dishes grouped.

It is better not to have a long menu as one competes with one's own dishes. Language should be simple and if possible effective Ascriptions should be given.

Important Factors that Influence Menu Planning are:

1. Money to be spent
   (a) customer.
   (b) food services.
2. Type of customer.
3. Type of food service.
4. Availability of foods.
5. Seasdnability of foods.
6. Balance.
7. Equipment.
8. Labour.
9. Leftovers in hand.
10. Appearance of the final product.
    (a) palatability (shape, flavour.)
    (b) variety (colour, texture.)
11. Menu-planning pattern.
12. Recipes to be used.
13. Food habits of the customer.
14. Policy of the establishment.

Nutritional Requirement

The nutritional needs of the people being catered for are an important consideration especially when planning set meals.

Nutritional needs differ for persons of various ages, sex and activity groups.

It is not necessary that we should have a balanced meal every time we eat. The day's meals together must be considered to see that the right quantity and quality are provided.

The primary consideration should be quality and not quantity. As a rule our appetites will give us a rough idea regarding quantity but when calculating quantities for institutional or large-scale catering, it is safer to calculate the total quantity, allowing 2400C (calories) per day per adult man, and using this unit to determine the amount required for women and children. The occupation of the people you are catering for must also be considered. 300 calories more per hour for very hard physical work, 150 to 300 calories for medium work and 75 calories for ordinary physical work should be added to the original amount. Allow about 150 calories per person per day for wastage.

No extra food is necessary for people with sedentary occupations. Mere variety is not enough. A knowledge of the different constituents of food and their functions in the body and the sources from which these foods are obtained should be studied.

Food Constituents

The different constituents of foods are carbohydrates, proteins, fats, minerals and vitamins. These are divided into:

(i) Body Warmers: These elements form the bulk of the food and are obtained from carbohydrates and fats. Although proteins can also be used for this purpose, it is wasteful to use a more expensive and less available food.

(ii) Body Builders: These are proteins and some minerals; meat, fish, milk and milk products, eggs and soybean are sources of complete or first-class proteins and therefore, are considered to be superior. Peas, beans, lentils and nuts are sources of second-class or partially complete proteins. A judicious mixture of these provides the best nutrition. Special stress must be laid on ensuring that when planning meals for vegetarians, a third of the proteins is obtained from the first category. When there is a shortage of first-class proteins it is a wise policy to distribute them though the three meals as proteins cannot be stored in the body. Minerals can be obtained from vegetables and fruits.

(iii) Body Protectors: These are vitamins, minerals and proteins. It is very easy to overlook the body protectors, particularly vitamins, as they are found in food in small quantities. So special attention must be paid to them. Cooking and processing destroy some of the vitamins, particularly Vitamin C. This vitamin is highly unstable and will be destroyed by oxidation, leaching, heat or ageing. Vitamin B is lost in the cooking liquor, if the liquor is thrown out, and destroyed when soda is used for cooking. Therefore it is essential that raw fruit, fresh vegetables, salads and some nuts or pulses be included in the daily diet.

Another way of dividing food is as follows:

(i) *For Bones and Teeth:* Calcium is necessary and it cannot be utilized without Vitamin D. Foods rich in calcium are milk and cheese; vegetables contain smaller quantities. Vitamin D is found in animal fat.

(ii) *For Flesh and Muscles:* These need proteins which are obtained from milk and milk products, meat, fish, poultry, eggs, soybean and a second-class variety from peas, beans, lentils, and nuts.

(iii) *For Blood:* This needs iron which besides giving red colour to the blood helps to transport oxygen to the various parts of the body.

(iv) *Food for Energy and Warmth*: Fat and carbohydrates. It is a good plan when planning meals to put in body-building and protective food first and then add other foods to satisfy hunger.

(v) *Food for Protection:* Such as minerals and vitamins. Vitamin A for growth, respiratory organs, skin and eyes. Vitamin B for healthy digestive system, nervous system and good appetite and for metabolism of carbohydrates. Vitamin C for cells of the body, clear skin, bright eyes and general good health. Vitamin D for fixing calcium and phosphorus in bones and teeth.

Food Planing for Children

While planning meals for children, besides a knowledge of their nutritional needs, the fact must be borne in mind that partaking of food is a significant social and educational experience. Food should be plain, well-cooked, easily chewed, free from strong flavours and odours. There should be interesting variety in colour, flavour and texture. Meals should be served well and meal-time should be pleasant, free from undue stress and resentment.

Aged people should have less calories (not more than 1500). Food should be easily chewed. Protective food such as vitamins and mineral should form the chief part. There should be less carbohydrates and fats. Foods served in commercial restaurants, cafeterias, industrial canteens and hotel dining rooms may constitute only a part of the daily consumption of foods. They should be planned to meet a third of the calorie requirement. Such institutions should so plan their meals that patrons will be able to readily select foods necessary for their bodily needs. This is facilitated by the arrangement of foods for a given meal into menus classed as light, medium and heavy. The inclusion of leafy vegetables and salad plates is a nutritionally sound principle.

Type of Service: Menus should be planned in relation to type of service e.g. cafeteria or seated service, etc., and the distance over which food must be transported for service. When meals are served in decentralized units from a central kitchen, a master menu with modifications should be used.

Availability and Seasonability: Seasonal foods are, besides being economical, also better in quality. These must be studied and included. It is also important to consider availability of food when planning menus.

Balance: The following points must be considered:

(i) Light to heavy and back to light,

(ii) Vary the sequence of preparation, e.g. cold or hot.

(iii) Change seasoning, flavouring and presentation,

(iv) Ensure that garnishes are in harmony with their main dishes.

Manpower

(i) Skill and number of employees must be considered,

(ii) Avoid too many last-minute processes,

(iii) Avoid too many long jobs especially for one cook,

(iv) Avoid overly complicated processes or emphasis on wrong kinds of food. Plan meals that the staff can do well.

(v) Have records of time and yield.

Leftovers: For wise and economic management, leftovers should be considered and the menu altered or the necessary adjustment made.

Look of the Final Product: Food should be palatable to be acceptable and hence recipes should be tried out and standardized before a dish is put on the menu. The tastes of the people catered to should be studied. New foods should be introduced gradually.

Natural colours should be used for food. Do not over-garnish. To make food attractive and to break monotony, variety in colour, texture and flavour must be introduced, e.g. Cream of tomato soup, Fish colbert with French fried potatoes/ Chicken fricassee with fresh peas and fruit salad or Puri and chole, Mughlai biryani, Palaksag, Tomato cucumber, Onion raita and Phren. As eye-appeal stimulates the appetite, relate the colours of the garnish with the basic ingredient.

Food Habits: This is important when catering in industrial canteens, schools, institutions, hospitals, etc. For example, in India, typically Jains and most Brahmins are strict vegetarians. Sikhs may eat only 'Jhatka meat'. Orthodox Jews may need Kosher meat. Bengal Brahmins may eat fish but not meat.

Policy of the Establishment: Certain establishments may not serve certain foods because of their policy, e.g. beef, or ham, bacon and pork. In order to have well-planned menus it is necessary to have a well worked-out menu pattern which is suitable to the clientele, and workable as it relates to staff and equipment.

What is a Well Planned Menu?

(i) Plan regular uninterrupted time for making your menus when you are not too fatigued. Choose a quiet place.

(ii) Have available all records, charts and books which will be helpful.

(iii) Have a well organized, classified and usable menu chart. This will be a guide to menu making and will, if kept up to date, suggest ideas and help to prevent monotony and repetition in menu making. Ideas for menu charts can be collected from books, magazines, visits to restaurants and many other sources.

(iv) Formulate a workable menu-pattern. This means that the menu-pattern should be planned to meet the needs of the various groups of individuals served, and should be based on definite information.

(v) Make menus one week in advance. Use menu charts and other sources to get interest, variety and nutritional adequacy, and to avoid repetition. Recheck menus daily to use up leftovers and to meet other problems.

Definitely, the arrangement of an elaborate dinner is an art in itself and consists of selecting the various courses so that the dishes harmonize with each other. The following points are helpful:

(i) Each dish should be different in composition and mode of cooking,

(ii) The choice of individual dishes must be excellent, each dish should be well cooked, tastefully dressed and yet distinct in character,

(iii) The harmonizing effect is obtained by so arranging the dishes that each one is distinct from the other, bearing no relation in appearance to the preceding or following dish,

(iv) A judicious selection of the raw materials with special regard to the season of the year must be made in all cases,

(v) The entree should always be classified so that light dishes such as rissoles, bouchees, croquettes, quenelles, chromesthesia, etc., are served first,

(vi) The various meats and materials forming one course must not be repeated in the same menu from one course to another,

(vii) When two or more types of fish are to be served the first should be boiled-- generally a large fish--and the second fried, grilled or broiled,

(viii) The various kinds of sauces used in the preparation and as the accompaniment must each be distinctly different in colour, taste and flavour,

(ix) If there are two soups, one ought to be clear and the other cream. If the cream soup is made of vegetables the clear soup must not contain any vegetables,

(x) If there are two releves, poultry is served before butcher's meat. The most important point is to have dressed releves before plain roasts. This is followed by the sweet, savoury, dessert and cheese. Sometimes the savoury is served before the sweet.

For writing menu use correct words and phrases. Choose the language easily understood by the type of client being served.

First-class establishments use the French language. Average establishments use English with a few common and easily understood foreign terms. Canteens use descriptive, easily understood English.

Do not mix languages. Always call an Irish stew by that name, or by the name used in the country of origin. Well known terms like hors d'oeuvre, etc., may be used when writing menus in English.

Some cooks give new names to old dishes, or give dishes names which have no bearing on the preparation.

This causes confusion and disappointment so this practice must be avoided.

When writing the menu it is very important to see that words are properly spelt. Call dishes by their proper names. Do not mask or disguise.

Use the following formula:

(i) Name the article--pomfret, lamb, duckling,

(ii) Add the place of origin--Bombay Pomfret, Punjab Lamb, Kerala Duckling,

(iii) Name the part used--Fillet of Bombay Pomfret, Saddle of Punjab Lamb, Breast of Kerala Duck,

(iv) Name the method of cooking--poached, roasted, braised.

(v) Add the garnish--in white wine, boulangre, and green peas,

(vi) Then having finished up with rather a long list of words, streamline the name leaving out the least important,

(vii) Finally add the sauce where necessary, Apple Sauce for the duck, Redcurrant Jelly for the saddle, etc.

What to Avoid?

(i) Never write the word "gravy" on the menu, (ii) Add after the main name any sauce which is served separately,

(iii) Add after the main name any accompaniment which is served separately viz: Yorkshire pudding, Cheese straws, etc.

(iv) Parsley is not put on the menu except where it is part of a dish, e.g. Parsley Potatoes.

(v) Watercress may be added either after the main word or by adding it with the word and.

(vi) Sauces and garnishes go directly after the dish to which they belong. Add the vegetable after the main dish and its garnish or sauce; and add the potatoes after the vegetable; add the salad after the roast or the cold dish.

(vii) Do not translate literally, e.g. Chasseur means hunter, Bonne Femme means good woman--in these cases either keep the original word or describe the dish as "with mushroom sauce" etc.

(viii) When something is grilled it cannot also be "a la meuniere". "A la" is only used to denote "in the fashion of, "after or in the style or. This means that it may be used when in the style of a certain town, country, place, etc., such as "a 1'anglaise", "a la francaise", "a la portugaise", etc. It might also mean "in the" as in "a la creme". But where a dish is named after somebody it will not have the words "a la" but go straight on to the name, e.g. Peach Melba. In the case of Vichy carrots the word "vichy" belongs to the water there and the correct term is, therefore, "Garrottes Vichy",

(ix) Do not translate words which are very well known like Mayonnaise, Hollandaise, Consomme, Yorkshire pudding, etc.

(x) Do not add fancy names not properly understood, like Grilled Plaice a la meuniere.

(xi) Do, not write 'soup and a roll': this looks clumsy.

(xii) Do not write Smoked Salmon Hollandaise. Salmon Hollandaise is hollandaise sauce mixed with poached fresh salmon.

(xiii) Do not repeat the garnish as a vegetable. The word 'Bruxelloise' means with brussels sprouts. Vichy means with carrots, Clamart means with peas, Boulangre means with-baked potatoes, and so on.

Correct Sequence of Courses: Make sure that the

sequence of courses is correct. Take as a sample those given later under respective headings. A few alterations are allowed sometimes; a savoury may be served before the sweet. The vegetable course is sometimes put before the roast. Cheese is sometimes served before the sweet.

Make sure that alternative dishes follow the same order of appearance as in a full menu, e.g. if it is fish or meat, then write the fish course first.

Correct Sequence of Garnishes and Accompaniments: First comes the main name of the dish, followed by the garnish which is to be served separately or the sauce which is served separately. Then the vegetables and then the potatoes.

Menu Card as a Tool

(i) Use good quality paper or paper mache or parchment or silk.

(ii) The print should be good and clear, in indelible ink, clean and elegant looking. Words should be correctly spelt. The prices should be up-to-date. A card with old prices crossed out looks untidy.

(iii) A medium-sized menu card is preferable to a too large or small one.

(iv) A menu card should have an attractive layout; one of topical interest is a good choice.

(v) There should be variety and contrast without confusing the customer.

Tea menus in hotels, cooked meals are seldom provided at teatime, other than boiled or poached eggs, but in commercial (public) restaurants the emphasis is on "high teas" and like conteens, they serve lighter snacks more on the lines of a snack bar.

Some Preferred Food

Breakfast	*Menu*		*Notes*
1st	Course: Fruit	a. Fresh grapefruit, melon, juices. b. Stewed compote of figs, prunes, apples, mixed fruit. Baked apple.	
2nd	Course: Cereals	Cornflakes, grapenuts, shredded wheat, etc., and always--porridge. Sometimes add the words--Hot or cold milk or cream.	
3rd	Course: Eggs	Scrambled, fried, boiled, poached, Scrambled, fried, boiled, poached, or back), tomatoes, mushrooms.	Where this is the main course, it will follow the fish.
4th	Course: Fish	Grilled mackerel, kippers, herrings, bloaters, fish cakes, haddock, kedgeree. Fried plaice sole, whiting, meuniere. Usually plain, sometimes with tartare sauce.	There are usually just single names.
5th	Course: Meats, Hot	Grilled bacon (back and streaky), ham, gammon, sausage, lamb's kidney, calves' liver, with sauteed or fried potatoes or potato fritters. Bubble and squeak, mushrooms, tomatoes.	
6th	Course: Meats, cold	Ham, tongue, pressed beef, rarely, poultry or butcher's meat; with plain or sauteed potatoes where not included above.	Salad may be served but not put on the menu.
7th	Course: Preserves	Marmalade, jam, honey.	
8th	Course: Fresh Fruit	Apples, pears, peaches, grapes, bananas, oranges, etc.	
9th	Course: Beverages	Coffee, tea, drinking chocolate.	
10th	Course: Breads	Rolls, crescents, toast, brioche, etc.	

Generally speaking, a breakfast will consist of three courses:

Fruit or cereal.

Eggs, fish or meat with or without potatoes.

Preserves and toast, rolls, coffee.

Sandwiches: Very small, dainty, white and brown bread, well buttered. Eggs and cress; tomato, lettuce, cucumber; chutney, savoury paste filling, sometimes chicken; foie gras (not often), smoked salmon (not often).

Hotplate Foods and Buns: Buttered buns, scones, teacakes, sally lunns, scotch pancakes. Doughnuts, waffles, samosas, sausage rolls, pakoras, mutton patties.

Bread and Butter: White, wholemeal, currant, fancy and proprietary brands.

Preserves: Jams, lemon curd, honey, etc. (not marmalade).

Toasted Items: Toast, teacakes, scones, crumpets, buns, etc. Cheese toast, farinaceous and savoury toasted items as in a snack bar.

Eggs: Boiled, fried and omelette.

Pastries: All varieties, afternoon tea pastries (small), gateaux, etc.

Ices: Plain ices: all varieties of sundaes.

Tea: Indian, Ceylon, China, Russian and iced.

Cold Dishes: Cold meats, poultry, foie gras, mousse, timbales, terrines, etc. with salad: most varieties.

3rd Course:
- (a) *Sweets:* Hot souffles, pancakes, fritters, etc. Cold iced souffle' bombes, biscuits, glace coupes, accompanied by petit fours.
- (b) *Savoury:* Most kinds are served.
- (c) *Cheese:* Most varieties.
- (d) *Dessert:* All the fresh fruits and nuts.
- (e) *Coffee:* Always.

Banquet Menus: (lunch) The usual lunch dishes are used (bearing in mind the suitability of service and keeping the dishes hot). It is not usual to include farinaceous dishes or eggs. Stews are avoided and savouries rarely served, but much may be made of cheese and biscuits. The menu again is shorter than its dinner equivalent and the service as a rule is fast and takes the following form.

1st Course: (a) *Cocktails:* Fruit, shellfish, smoked fish,
(b) *Hors d'oeuvre,* or
(c) *Soups.*

2nd Course: *Fish.*

3rd Course: *Entree:* (excluding stews and rechauffe dishes) but including dinner entrees, roast joints, grills or cold meats. With vegetables and potatoes or salads and potatoes.

4th Course: *Vegetable dish:* Asparagus if in season.

5th Course: (a) *Sweets:* With emphasis on the hot variety including bombes, biscuit glace as at dinner.
(b) *Cheeses:* Most varieties with biscuits and celery or radishes.
(c) *Dessert:* Rarely.
(d) *Coffee:* Always.

Light Buffets

Light buffets may be required for all kinds of catering--to augment a long function (dance) some hours after the main meal, for tea dances, supper dances and the like. The light buffet can be of two sorts, an inclusive "help yourself" running buffet--where the food is displayed for everyone to choose--or an a la carte buffet which is served by waiters and not usually displayed. The a la carte will follow the same lines as a displayed buffet but may include less of the smaller items and more of those prepared to order in the kitchen. The general arrangements will be a boxed-in line of tables, where guests can walk around with enough space inside for the servers. The a la carte buffet will include a menu card (priced) and waiter service.

The menu should consist of items easily eaten with the fingers, though forks and servilities will be provided. In connection with this kind of work one often hears of a "Fork Luncheon" or "Fork Buffet" and the idea behind this means that guests will probably be standing (that is, tables will not be provided nor chairs) and food will be any items easily eaten with a single implement (fork or spoon) while the guest is standing with the plate in one hand and the fork in the other.

Most of the following light buffet items will be suitable for " Fork" parties even with the addition of anything soft like chicken mousse, galantine, terrines, chicken and other salads, etc.

Light Buffet Menu

Savoury Finger Foods: To include any of the cold canapes--chipolata sausages on sticks; wrapped in bacon; prune sticks; celery branches piped with cheese spread; game chips, or gaufrette, very dry and seasoned; patties of chicken, lobster, crab, salmon; sausage rolls and any other puff pastry savoury snack; sandwiches: cut very small, brown or white bread --smoked salmon, egg and cress, tomato, ham, tongue, chicken, etc.

Sweets: Charlotte Russe, jellies, trifles, bavaroise, etc. fruit salads or fresh raw fruits like strawberries. Ice creams, various flavours, but not as a rule coupes. Pastries, gateaux, biscuits

Beverages: Coffee, tea, punch, iced coffee.

Supper Buffets: Generally speaking, a Supper Ball will begin late and a full buffet will be available. This is not served course by course, but will be rightly divided by lines or spaces in the printed menu denoting the different types of food as follows:

Shellfish Oysters, prawns.

Smoked Fish and Meat: Smoked salmon, ham, trout, eels, etc.

Canapes: Savoury finger toasts, with various spreads.

Savouries Chicken: lobster, crab, salmon patties, game chips, chipolata, celery sticks, cheese straws.

Fish: Cold salmon, lobster, crab, fillets of sole, pomfret in aspic.

Meals: Chicken supremes, mayonnaise and salads; turkey roast--cold; ham--cold, decorated and jellied; mousse and cornets; cutlets in aspic; galantine, foie gras in aspic.

Salads: Lettuce, Russian and all kinds. Mixed and plain. *Sandwiches:* Ham, tongue, chicken, turkey, smoked salmon, etc. *Bread Rolls:* Plain, egg and cress, etc., milk rolls.

Cold Sweets: Charlottes, jelly trifle, bavaroise, cream, caramel, condes, flans. French pastries, various gteaux.

Ices: Various bombes, biscuits, glaces with fruit and petit fours. *Beverages:* Coffee, tea, punch, iced coffee and tea.

Soups: Sometimes hot turtle soup or a cold consomme (madrilene) will also be served at the commencement of the buffet, or it may be served at the end of the ball in which case it will be printed "en partant".

All the dishes will be very well decorated, making a great display, and quite often the ball will finish up with a breakfast dish such as eggs and bacon, kippers, etc., or as mentioned before, a bowl of soup such as: hot double consomme (clarified twice) or turtle soup, or very often soupe a l'oignon.

Outdoor Catering

Utmost care must be taken in selecting the menu, keeping in mind the facilities available on the spot, and what must be taken to augment them. The secret of outdoor catering is proper equipment, properly equipped vans, trained staff, and a knowledge of weather conditions and the number likely to be fed. Among the equipment will be shamans, marquees, trestle tables, chairs, containers, calor gas equipment, trays, wash-up facilities, guarded arrangements (where necessary) for the payment of money, as well as the usual equipment.

Weddings Parties: Wedding parties follow either the ordinary sit-down meal pattern, or, more often than not, have a running buffet. The wedding cake will always be the centerpiece, and more floral displays will be arranged than usual. The flowers of the bride and bridesmaids may be laid alongside the cake. A silver salver will be ready to receive the telegrams. Particular care will be taken with the seating plan in conjunction with the parties. For the reception, sherries may be served (or cocktails) while the guests are congratulating the married couple, or sometimes after the nuptial service (which the guests may have attended) a cup of tea and biscuits may be served.

1st *Course:* Cocktails, smoked salmon, fruit, hors d'oeuvre or a good soup. :

2nd *Course:* Cold salmon, roast poultry, grilled butcher's meat, cold supreme of chicken, aspics, etc., with vegetables and potatoes or salad and potatoes.

3rd *Course:* A very good sweet course; fruit sundaes, coupes, biscuits, bombes, with most likely an alternative of the cold sweets, flans, trifles, etc., served with fruit, and sometimes with petit fours or a course of French pastries.

4th *Course:* Coffee.

The cake is then cut and handed round, and the bride and groom are toasted with champagne. Then follow the reading of the telegrams and the speeches.

In buffet wedding receptions several dishes will be chosen from the Supper Ball menu-- leaving out the hot soup, with fewer savouries and sandwiches, but a full range of cold meats--or a light buffet may be chosen as sufficient to meet the need, and tea will be served instead of coffee.

Cocktail Parties: Apart from the actual drinks which are the main item for a cocktail party, very tiny snacks (savoury), cigarettes and a well equipped bar are the other necessities.

The food will be dressed attractively, but will also be handed round with the cocktails by the waiters. The menu (never printed or shown) can be all kinds of savoury finger toasts, cocktail olives, stuffed olives, gherkins, walnuts, pickled onions, etc., very crisp, salted game chips, salted almonds, chipolata sausages, bacon rolls, prune and bacon rolls, cherry and bacon rolls, all on sticks. Stuffed sticks of celery, cheese fritters, cheese straws, very tiny savoury filled sandwiches. The fillings used for the canapes are caviar, smoked salmon, smoked ham, cooked ham, anchovies, sardines, shrimps, freshly made savoury purees for spreading, sliced egg, cheese, etc. Some of them will be spread on crisp toast, and others on thin, crisp flaky pastry or biscuits.

Hospital Catering: Nowadays, this is better organised than it was a some years ago--diets are in the charge of a dietitian, and catering is supervised by a catering officer. Their three main functions are to feed the staff, patients, and visitors. Staff will be provided with good canteen food or, as is most often the case, with the ordinary hospital diet, and the canteen will stock light refreshments. The food leaves the main kitchen in electric hotplate trolleys, and is sent to each ward which has an anteroom with an adjoining 'mise-en-place' of small equipment for making tea, bread and butter, light items like toast, etc., and salads. Each ward sends an order daily for the three meals with requests for full, light, fluid or special diets. The tea meal is usually prepared in this service room.

The three meals ordered from the kitchen will be breakfast, lunch, and either supper or dinner.

Normal diets, or Full Diets: Will be well balanced normal meals of protein, carbohydrates (but with very little fatty food), vitamins, mineral salts, with emphasis on salads, fresh fruits, few spices and never shellfish or rechauffe dishes.

Breakfast: Fresh fruit juices, grapefruit, fresh fruit, cereals and porridge.

Fish: Haddock, kippers, herrings, fish cakes. *Eggs:* All kinds.

Meats: Bacon, sausage, rarely other meats. Vegetables: Confined to tomatoes, rarely potatoes. Preserves, toast and beverages as usual.

Lunch: Fresh fruit juices, fresh fruit.

Soups: Particularly consomme and broths.

Fish: Deep-poached, or steamed with simple sauces, sometimes grilled and fried.

Entrees: Stews, boiled chicken and meats, pies and hot pots, liver, heart, or roast joints. Vegetables: Fresh and seasonable variety. Potatoes: All kinds except the fatty ones. Cold meats with fresh salads. Sweets: Milk puddings, sponge puddings, blancmange, jellies, stewed fruit, tarts, ices. Cheese: Sometimes.

Tea: Will follow a very light pattern.

Supper or dinner will follow the same pattern as for lunch but will be lighter in substance often with a farinaceous dish, cheese, egg, stuffed marrow, soup, salad, soused herring, cauliflower, sausage rolls, or the like instead of the main meat course.

Light Diets: Selected light dishes will be chosen from the ordinary menu and supplemented with plenty of egg dishes, broths, jellies, junket and fruit such as those shown under "Invalid Cookery".

Invalid Cookery: Dishes should be made to look very tempting, dainty, colourful and inciting to a poor appetite.

Breakfast: Milk cereals, flaked haddock, steamed fish, eggs, herring roe. Other meals:

Soups: Consomme and all the broths (sometimes strained).

Eggs: Lightly boiled, scrambled, poached and omelettes.

Fish: Steamed and boiled white fish, pie, and purees.

Meats: Boiled chicken, turkey, lamb, sweetbreads thinly sliced, minced chicken, turkey, lamb, beef, liver, steamed lamb cutlets. Cheese dishes: Custards, semolina, welsh rare bit. Vegetables: Boiled fresh vegetables, usually pureed, never fried, sauteed or braised. Salads and potatoes, usually pureed, never fried or sauteed. Cold meats: Ham, tongue, thinly sliced.

Sweets: All kinds of trifles, jellies, milk jellies, junket, egg custards, purees of fruit, baked fruit, fools, ices.

Liquid Diets: Milk, soda water, fresh fruit juices, consomme, beef tea, broths and glucose water.

Special diets: Special diets will be the job of a certified dietitian or the doctor and should never be selected by the caterer. Special breads, sugar, jams, starches, etc., will be supplied for diabetics.

Special care should be opted in buying for invalids or hospitals. Only the freshest of vegetables, finest of fish and meat, newly laid eggs, pasteurized milk, etc., should be used in the kitchen.

School Meals

In case of schools Here, proper knowledge and practice of well balanced meals is expected--body-building, nourishing, sustaining and full of vitamin content. The practice of central kitchens is gradually being superseded by kitchens on the spot. Insulated containers from central kitchens were never successful as the food never looked, smelled or tasted appetising. Great stress is laid upon "per head" quantities and "per head" costing. Diet sheets are made out in advance by school meals supervisors. Well balanced meals cooked to retain as much nourishment and flavour as possible should be stressed.

Snack Bar and Milk Bar: Three prime requisites--either a bar with stools or tables and chairs; tea and coffee-making machines and a soda bar space for the dispensing of icecream and soft drinks; as well as small equipment and other devices necessary for speedy service. The menu will contain light snacks: hot soups, beans, spaghetti, etc., on toast, salads and meat salads, sandwiches and filled rolls, cakes, buns and pastries. Great emphasis is given to ice-cream sundaes and soft drinks as well as tea and coffee and all kinds of milk shakes, Bovril, Horlicks, and in fact all those items called "counter lines".

Seasonal Variations in Diet: Planning preparation and service of meals, be it for the family or for institutional feeding, should be based on certain fundamental principles such as nutritional requirements, money to be spent, palatability, seasonability and availability, variety and appearance of material, under which will come variety in colour, flavour, etc.

One of the most important aspects of meal planning is 'seasonal variations in diet'. While quite a number of the other principles of meal planning are based on this, very little thought and attention are given to this aspect.

Seasonal variation in diet has two interesting divisions:

1. Seasonability and availability of raw material.
2. Variations in the type of dishes depending on climatic changes.

Food should be eaten as far as possible when it is at its best. When any food is in season it is at its best nutritionally, aesthetically and from the palatability point of view, apart from the fact that it can be fitted in more easily into the budget. In other words, 'Value for money' should be the aim of every caterer. Remember, however, that it is poor economy to buy goods of inferior quality because they appear to be cheap. There is a difference between buying cheap and inferior foods and buying good quality food at cheaper rates because it is in season. Among the tools that a food service manager should have to help him plan meals wisely should be a systematically arranged chart showing the seasonability of fruits, vegetables, meat and fish. In Mumbai we could organize this chart in two ways. One way is to have 3 sections:

(a)	Monsoon	June to October
(b)	Winter	November to March
(c)	Summer	March to June

Under each column you could start with vegetables, then fruit (with peak periods in brackets) and finally fish, meat and eggs.

Fish is available from immediately after the monsoon, about the middle of September, to June. It is fresher and tastier in the winter months.

MONSOON	*WINTER*	*SUMMER*
June-Oct.	Nov.-March	April-May
Vegetables:	*Vegetables:*	*Vegetables:*
Green leafy vegetables	Beetroot	Cabbage
Brinjal	Carrot	Chowli beans
Ridge grourd	Cauliflower	Round gourd
Snake gourd	Capsicum	Gherkin
Corn	Celery	Spinach
Radish	French beans	Yam
Ladies fingers or okra	Tomatoes	Snake gourd
Sweet potatoes	Peas	Ash gourd
Double beans	Parsley	Drumstick
Sword beans	Salad leaves	Garlic
Colocasia	Turnips	Onions
Ginger	Kohlrabi	Vegetable marrow
Green chillies	Leek	Cucumber
	Lime	Colocasia
Fruits:	*Fruits:*	*Fruits:*
Sweet limes (Oct.)	Mandarin orange	Melon
Bananas	Sweet lime	mango
Limes	(Feb.-March)	pomello
Jammu plums	Grapefruit	Strawberry
Jammu apricot	Apple	Figs
Jammu peach	Pear	Grapefruit
Mango	Nasik grapes (Feb.-May)	Lime
Chikkoo	Pomegranate	Jammu Plum
Pineapple	Custard apple	Banana
Apple (Sept.-Oct.)	Loquat (Feb.-March)	Chikkoo
Mandarin orange	Wood apple	Surat guava
		Jackfruit
		Palmyra kerne
		Tender coconut

The price of meat generally varies little with the season. Pork however, is cheaper from September to April, perhaps because it is susceptible to deterioration at higher temperature. Being a meat which has plenty of fat it is best avoided during the summer months. Eggs keep-better during the winter months and hence are cheaper as well.

Another form of dividing foods could be to classify them item-wise, giving the source of supply, season and peak period. For example, mangoes from the Konkan area are available from March to the first week in June, with a peak period of April and May, and the ones from Gujarat are available from May to June.

If feasible market trends must be studied thoroughly to know what is available, and to keep in touch with current prices. Experience has the greatest value but the inexperienced can learn by studying the market quotations published weekly in the papers and by actually studying the market.

Equally important is the second aspect--variations in the type of dishes depending on climatic changes. During the summer months appetites are on the decline and morale rather low. Greater care and attention has to be given to planning of meals to whet the appetite, to provide an interest in food and to maintain general health. Food in summer should be light, attractive, easy to prepare, and freshly prepared. Plenty of liquids and fruits should be included. Avoid spicy and greasy foods. Lime juice, either freshly prepared or prepared and stored in the refrigerator for not more than a week, to retain vitamin C is a must.

Melon cocktails (scoop out centres, remove seeds, mix with sugar and lime juice), buttermilk flavoured lightly with green chillies (without seeds) and onions with salt added, mango fool (cook mature but unripe mangoes with sugar, mash and mix with custard, chill and serve) are all welcome changes during summer. Ice-cream and water ices or ice-fruits are great favourites with children. Scoops of ice-cream could be mixed with lemonade to make a tall fizzy drink for the grown-ups.

Cold coffee and iced lemon tea could be an interesting change. Salads, particularly cucumber and cabbage, as both these vegetables are in season could be included in the menu. A French dressing or a dressing of lime juice, olive oil and salt will be preferable to rich mayonnaise or other salad dressing. A word of caution--cabbage should be shredded very fine when eaten raw, so that it does not cause indigestion and flatulence.

Varieties of cold meats and salads, or cold fish with a tartare or mayonnaise sauce and salad, could make pleasant luncheon menus for those who like western preparations. Rich pilaus and biryanis, fried rice, etc., should be given a miss. A change from boiled rice could be had by having curd bath, lime rice, etc. Cucumbers, salads anctraitas must be included. During the monsoon, hot snacks and savouries can be introduced. Corn on the cob is another seasonal vegetable that could be introduced either roasted over a charcoal brazier (sigree) or boiled and buttered. Most of the so-called Indian vegetables are available during the rains. Some of the traditional Indian dishes such as spicy brinjal bhath, kadambom, masala stuffed ladies fingers, masala bhara mirchee, etc., could be included to provide variety.

With the abundance of fruits and vegetables available in the winter months there are plenty of opportunities to indulge in creative and interesting culinary preparations. There is no reason why this should mean expensive foods. It is more an opportunity of providing variety in colour, texture and flavor. Fish is at its best during this season. This is also the time when rich foods like biryanis and moghlai dishes are popular. See that foods that provide the essential vitamins and minerals are included to keep off the common cold and cough.

Knowing what is available in the market so that you can get the best for your money and introducing dishes that suit climatic conditions. Wise and intelligent planning and an interest in food are requisites.

Low-Calorie Diets

Very few people are overweight because of bodily disturbances. Most people are fat because they are sedentary in their occupation but eat as if they are vigorously active. Food, always a symbol of hospitality and good fellowship, is becoming more and more important. Alone or with others, most people love food. It reduces boredom and in times of stress and strain it acts as a palliative. When unhappiness is prolonged people can develop habits of over-eating that last a lifetime.

No one weight is right for all people of the same height, sex and age. Ideal weight depends in large measure on bone structure. Some people are short and stocky while others are tall and slender. Some have large bones and some have small bones. The following chart gives the desirable weights for men and women over 25. Weights are given according to frame, as normally dressed. After checking the weight scale it is desirable to plan the daily calorie allowance.

Science has found that moderately active adults require about 33 calories for each kg. (15 calories for each pound) of body weight, or less if they are not so active. For manual labour, allow 44 calories for each kg. (20 calories for each pound) of ideal weight. For safe weight reduction, many doctors limit weight loss to a maximum of 1 kg or 2.2 pounds a week. A man who normally eats about 2500 calories worth of food daily can cut 1000 calories from his diet and still get all the proteins, vitamins and minerals he needs. A woman who normally eats only 2000 calories worth of food can spare only 500 to 700 calories each day, which means a somewhat slower weight loss. Reducing however, should be done gradually, working from a 2000-calorie diet to a 1000-calorie diet, and should stop at a point when the person is able to perform his usual work without feeling any weakness or general debility.

Remember and keeps in mind the following factors:

1. Less food, but not so very much less as to affect health.
2. Avoid too much variety and spices.
3. Avoid sugar and sweets.
4. Use less fat and avoid rich meals.
5. Milk should be taken in moderation.
6. No puddings and pastries.
7. Use plenty of green vegetables but a reduced quantity of rice and chapatties. Avoid root vegetables.

2000-CALORIE DIET *(Non-Vegetarian)*

Breakfast:	
Bread	60 gm (2 oz) 2 thin slices
Egg	1
Milk	235 ml (8 oz)
Coffee or tea as desired, without sugar	
Mid-morning:	
Milk	175 ml (6 oz)
Noon:	
Rice	85 gm (3 oz) (225 gm or 9 oz cooked weight)
Dal	30 gm (1 oz) (85 gm or 3 oz cooked weight)
Mutton	85 gm (3 oz)
Curds	85 gm (3 oz)
Vegetables (not root)	1 helping
Greens	1 helping
At 4 p.m.:	
Milk	235 ml (8 oz)
Coffee or tea as desired, without sugar.	
Fruit (cirrus)	1 portion

Dinner:

Rice	60 gm (2 oz) (170 gm or 6 oz cooked weight)
Dal	30 gm (1 oz) (85 gm or 3 oz cooked weight)
Mutton	60 gm (2 oz)
Vegetables (not root)	1 helping
Greens	1 helping
Milk	235 ml (8 oz)
Daily allowance of oil in cooking	15 gm or V2 oz

Summary of Diet:

Carbohydrates	249 gm
Proteins	91 gm
Fat	70 gm
Calories	1996

N.B.: Skimmed milk or toned milk may be used instead of whole milk, and corresponding portions of butter could be used.

2000-CALORIE DIET *(Vegetarian)*

Breakfast:

Bread	60 gm (2 oz) 2 thin slices
Milk	235 ml (8 oz)
Coffee or tea as desired, without sugar	
Banana	1

Mid-morning:

Milk	175 ml (6 oz)

Lunch:

Rice	85 gm (3 oz) (225 gm or 9 oz cooked weight)
Dal	45 gm ($1^1/_2$ oz) (130 gm or 41/2 oz cooked weight)
Curd	85 gm
Vegetable (not root)	1 helping
Greens	1 helping

At 4 p.m.:

Milk	235 ml (8 ozs.)

Banana	1

Coffee or tea as desired, without sugar.

Dinner:

Chapatties	60 gms. wholewheat flour (2 oz) (3 phulkas weighing 20 gms each, 12.5 cm. (5") in diameter)
Dal	45 gm (($1^1/_2$ oz) (130 gm or 41/2 oz cooked weight)
Greens	1 helping
Vegetables	1 helping
Milk	235 ml (8 oz)
Daily allowance of butter or oil for cooking	30 gm

Summary of Diet:

Carbohydrates	291 gm
Protein	72 gm
Fat	62 gm
Calories	2013

2000-CALORIE DIET *(European Non-Vegetarian)*

Breakfast:

Porridge (oatmeal, semolina or dalya)	(15 gm $^1/_2$ oz)
Milk	
Egg	1
Bread	60 gm (2thin slice)
Banana	1

Mid-morning:

Meat extract or unsweetened lime juice	1 cup

Lunch;

Clear soup	150 ml (5 02)
Mutton	85 gm (3 02)
Potatoes	115 gm (4 oz)
Salad vegetables or greens	1 helping
Milk	150 ml
Fruit	1 portion

At 4 p.m.:

Bread	60 gm (2 thin slices)
Butter	15 gm ($^1/_2$ oz)
or Cheese	il0 gm ($^1/_2$ oz)
Milk	120 ml (4 oz)
Coffee or tea as desired,	without sugar.

Dinner:

Clear soup or strained vegetable soup	150 ml (5 oz)
Mutton	85 gm (3 oz)
Potatoes	115 gm (4 oz)
Greens or salad	1 helping
Milk	150 ml (5 oz)
Fruit	1 portion

Bedtime:

Milk	235 ml (8 *oz)*

Summary of Diet;

Carbohydrates	216 gm
Protein	87 gm
Fat	83 gm
Calories	1954

1500-Calorie Diet *(Non-Vegetarian)*

Breakfast:

Bread	60 gm (2 thin slices)
Egg	1
Milk	170 ml (6 oz)

Lunch:

Rice (255 gm or 9 oz cooked weight)	85 gm (3 oz)
Dal (130 gm or IVg oz cooked weight)	45 gm ($^1/_2$ oz)
Curd	85 gm
Vegetables (not root)	1 helping
Greens	1 helping

At 4 p.m.:

Milk	235 ml (8 oz)

Coffee or tea as desired, without sugar.

Fruit	1 portion

Dinner:

Chapaties	60 gm wholewheat flour (2 oz)
(3 phulkas weighting 20 *gra* each)	(12.5 cm. (5") in diameter).
Dal	45 gm $^1/_2$ oz)
Curd	85 gm (3 oz)
Vegetables (not root)	1 helping
Greens	1 helping
Daily allowance of fat	10 gm (*V$* oz)

If desired skimmed milk can be taken, and the extra fat used for cooking purposes.

Summary of Diet:

Carbohydrates	259 gm
Protein	60 gm
Fat	27.5 gm
Calories	1528

1500-CALORIE DIET ***(European Non-Vegetarian)***

Breakfast:

Porridge	15 gm ($^1/_2$ oz)
(oatmeal, semolina or dalya)	
Milk	150 ml (5 oz)
Toast	60 gm (2 oz) (2 thin slices)
Butter	15 gm (½oz)
Egg	1

Mid-morning:

Meat extract or 1 cup unsweetened lime juice.

Lunch:

Clear soup or strained	
vegetable soup	150 ml (5 oz)
Mutton	85 gm (3 oz)
Potatoes	60 gm (2 oz) (1 medium sized)
Vegetables (not root)	1 helping
Salad	1 helping
Fruit	1 helping

At 4 p.m.:	
Bread	3 thin slices
Butter	15 gm (Vg oz)
Tomato or lettuce	
Milk	150 ml (5 oz)
Coffee or tea as desired,	without sugar.

Dinner:	
Clear soup or strained vegetable soup	150 ml (5 oz)
Mutton	85 gm (3 oz)
Potatoes	60 gm (2 oz)
Vegetables (not root)	1 helping
Greens	1 helping
Fruits	1 portion

Summary of Diet:	
Carbohydrates	168 gm
Protein	64.5 gm
Fat	61 gm
Calories	1483

1200-CALORIE DIET *(Non-Vegetarian)*

Breakfast:	
Bread	60 em (2 oz) (2 thin slices)
Egg	1
Milk	170 ml (6 oz)

Coffee or tea as desired, without sugar.

Mid-morning:	
Unsweetened lime juice	1 cup

Lunch:	
Rice	60 gm (2 *oz)* (180 gm or 6 oz cooked weight)
Mutton	60 gm (2 oz)
Curd	85 gm (3 oz)
Vegetables (not root)	1 helping
Greens	1 helping

At 4 p.m.:	
Milk	120 ml (4 oz)

Coffee or tea as desired, without sugar.

Fruit	1 portion
Dinner;	
Mutton	60gm (2 oz)
Vegetables	1 helping
Chapaties	60 gm (2 oz) wholewheat flour (3 phulkas of 20 gm each 12.5 cm. (5) in diameter)
Daily allowance of fat in cooking	10 gm (1/3 oz)
Summary of Diet:	
Carbohydrates	166gm
Proteins	52.5 gm
Fats	39 gm
Calories	1222

1200-CALORIE DIET *(European Non-Vegetarian)*

Breakfast:

Bread	60 gm (2 oz) (2 thin slices)
Butter	7 gm (1/4 oz)
Egg	1
Milk	90 ml (3 oz)

Coffee or tea as desired, without sugar.

Fruit	1 portion

Mid-morning:

Meat extract or 1 cup unsweetened lime juice

Lunch:

Clear soup or strained vegetable broth	150 ml (5 oz)
Mutton	85 gm (3 oz)
Potatoes	60 gm (2 oz)
Vegetables	1 helping
Greens or salad	1 helping
Fruit	1 portion

At 4 p.m.:

Milk	60 gm (2 oz)

Coffee or tea as desired, without sugar.

Bread	60 gm (2 oz)
Butter	7 gm (1/4 oz)
Dinner:	
Clear soup or strained broth	150 ml (5 oz)
Potatoes	60 gm (2 oz)
Vegetables (not root)	1 helping
Greens	1 helping
Milk	150 ml
Egg	1 custard without sugar
Summary of Diet:	
Carbohydrates	134 gm
Proteins	60.5 gm
Fats	49.5 gm
Calories	1221

1000-CALORIE DIET *(Non-Vegetarian)*

Breakfast:	
Bread	30 *gm* (1 oz) (1 slice)
Egg	1
Milk	170 ml (6 oz)
Coffee or tea as desired, without sugar.	
Mid-morning:	
Unsweetened lime juice	1 cup
Lunch:	
Rice	45 gm (1 1/2 oz) (130 gm or 4 1/2 oz cooked weight)
Mutton	115 gm (4 oz)
Curd	85 gm (3 oz)
Vegetables	1 helping
Greens	1 helping
At 4 p.m.:	
Milk	120 ml
Coffee or tea as desired, without sugar.	
Dinner:	
Rice	30 gm (1 6z) (90 gm or 3 oz cooked weight)

Egg	1
Vegetables	1 helping
Greens	1 helping
Daily allowance of fat in cooking	(10 gm 0/3 oz)

Summary of Diet:

Carbohydrates	104 gm
Protein	51 gm
Fat	42.5 gm
Calories	1000

1000-CALORIE DIET *(Vegetarian) Breakfast:*

Bread	30 gm (1 oz) (1 slice)
Fruit	1 portion
Milk	170 ml (6 oz)

Coffee or tea as desired, without sugar.

Mid-morning:

Milk	170 ml (6 oz)

Coffee or tea as desired, without sugar.

Lunch:

Rice	30 gm (1 oz)
	(90 gm or 3 oz cooked weight)
Dal	30 gm (1 oz)
	(90 gm or 3 oz cooked weight)
Vegetables	1 helping
Greens	1 helping

At 4 p.m.:

Milk	170 ml (6 oz)

Coffee or tea as desired, without sugar.

Dinner:

Rice	30 gm (1 oz)
or	(90 gm or 3 oz cooked weight)
Chapaties	30 gm wholewheat flour (1 oz)
Dal	30 gm (1 oz)
	(90 gm or 3 oz cooked weight)

Curds	85 gm (3 oz)
Greens	1 helping

Daily allowance of 10 gm (1/3 oz) fat in cooking

Summary of Diet:

Carbohydrates	152 gm
Proteins	47 gm
Fats	29.5 gm
Calories	1061

1000-CALORIE DIET *(European Non-Vegetarian)*

Breakfast:

Bread	60 gm (2 oz) (2 thin slices)
Butter	15 gm (1/2 oz)
Eggs	2
Milk	120 ml (4 oz)

Coffee or tea as desired, without sugar.

Mid-morning:

Meat extract or 1 cup of unsweetened lime juice.

Lunch:

Clear soup	150 ml (5 oz)
Mutton	85 gm (3 oz)
Vegetables	1 helping (not root)
Greens or Salad	1 helping
Fruit	1 portion

At 4 p.m.:

Milk	120 ml (4 oz)

Coffee or tea as desired, without sugar.

Dinner:

Clear soup	150 ml (5 oz)
Fried fish or	85 gm (3 oz)
Steamed fish	145 gm (5 oz)
Potatoes	60 gm (2 oz)
Vegetables	1 helping
Greens	1 helping
Fruit	1 portion

Milk	120 ml

Coffee or tea as desired, without sugar.

Summary of Diet:

Carbohydrates	95 gm
Protein	55 gm
Fat	51 gm
Calories	1054

ALTERNATIVES FOR USE WITH WEIGHED DIETS

One Portion of Fruit Yielding 10 Gm of Carbohydrates

Apple (medium-sized fruit)	85 gm (3 oz)
Banana (1 small or 1/2 big)	60 gm (2 oz)
Figs (fresh)	60 gm (2 oz)
Grapes	85 gm (3 oz)
Grapefruit	145 gm (5 oz)
Pomelo	145 gm (5 oz)
Guava	115 gm (4 oz)
Jackfruit	60 gm (2 oz)
Ripe mango (medium size)	85 gm (3 oz)
Yellow melon (½ fruit)	170 gm (6 oz)
Watermelon	285 gm (10 oz)
Orange or sweet lime (medium-sized fruit)	115 gm (4 oz)
Papaya	115 gm (4 oz)
Peaches	145 gm (5 oz)
Chikkoo	50 gm (2 oz)
Tree tomato	115 gm (4 oz)

Substitute 60 gm (2 oz) of bread with one of the following:

1 idli

1 appam

1 dosa prepared from 30 gm cereal and a little chutney.

upuma

Substitute 60 gm rice with

Bread

Chapaties prepared from	60 gm wholewheat flour (3 Rots of 20 gm each and 12.5 cm. (5") in diameter)

Substitute 85 gm (3 oz) mutton with:

liver	85 gm (3 oz) or

fried fish	85 gm (3 oz) or
steamed or boiled fish	145 gm (5 oz) or
fried brain	85 gm (3 oz) or
cheese	45 gm ($1^{1}/_{2}$ oz) or
chicken	45 gm (1/3 oz) or
boiled brain	145 gm (5 oz)

One portion of vegetables may be: green or leafy vegetables, carrot, onion, brinjal (egg plant), drumsticks, kohlrabi, tomato, vegetable marrow, gourds, pumpkins, cauliflower, ladies fingers, green beans. Avoid peas, beans and pulses, if not specially mentioned.

The still-room or pantry is an all-purpose department and is open as long as service of food and beverages is required. Its chief function is the preparation of sundry foods and beverages. It is, therefore, necessary that all the staff who work in this department are properly trained. This is necessary on two counts, firstly to see that the food and beverages are skilfully prepared, and secondly to see that the portioning is done properly, as the rate of gross profit earned in this department should be higher than the other departments.

The following are some of the foods and beverages required.

Breakfast: All beverages, cereals, fruit juices, fruit, toast, butter, preserves, etc.

Mid-morning: Coffee, tea, snacks, etc.

Lunch: Beverages including minerals and squashes, toast melba, salads, etc.

Afternoon Tea: Complete service.

Dinner: Beverages, sundries, finger bowls, etc.

Obviously, it will be seen that it is essential to have a tight control system operating here, otherwise the gross profit will not be earned. The setting is important in kitchen planning and dining room service. It is necessary to have the still-room adjacent to the dining-room and with easy access to the kitchen.

It is not advisable to make the still-room too small, as space must be made for the following equipment:

(i) A cafe set comprising counter-type equipment for coffee, tea and milk urns.

(ii) Hot water boiler,

(iii) Self-timing toaster,

(iv) Bread and butter machine if large tea trade is done,

(v) A tea-dispenser to ensure accurate measurement of portions, thus avoiding waste,

(vi) A butter-pat machine ensures quality of portions and aids costing,

(vii) A coffee percolator or Espresso machine according to the size of demand,

(viii) Glass-washing machine so that maximum hygiene can be practised,

(ix) Two-sink (stainless steel) system for washing crockery.

(x) Linen basket for dirty linen,

(xi) Storage cupboards,

(xii) Brain-Marie and hot closet for the service of hot dishes, sauces, soups, etc.

(xiii) Some establishments prefer to grind coffee as it is required. For this a small grinder is necessary.

It will readily be seen that proper control of the still-room is vital to the successful operation of the food management of a hotel or catering establishment.

Buffet Service

Buffet service is again becoming increasingly popular.

It was the custom years ago to have a large collection of dishes, both savory and sweet, available on a buffet for breakfast, luncheon and supper.

A buffet for breakfast is no longer in vogue as much simpler menus are used for this meal.

For lunch and supper and even dinner, the number of dishes served from a buffet are again increasing. In addition to a large variety of Indian curries, savouries and sweets, typical Western cold collations are included. A cold buffet can contain as much as 40 to 50 dishes.

1. Hors d'oeuvre: caviar, pate de foie gras, fruit cocktail, fruit juices. Lobster mayonnaise, salmon mayonnaise, aspic of crab, prawn, shrimp, scampi.
2. Cold consomme, cold salmon, fish mayonnaise, cold meats such as pressed spiced meat, brisket of beef, silverside of beef, spiced glazed tongue, roast meat, roast poultry, raised pies.
3. Prepared salads, Russian, French, Italian, etc., and separate salad material-- cucumber, beetroot; mustard, vinegar, etc., for the customer to make his own dressings; salad dressings, mayonnaise; sauces such as Horseradish, Cumberland, mint; proprietary pickles and chutneys.
4. Game pie, mushroom and oyster pie, cold poultry, spring chicken, capon, cold game such as partridge, pheasant, grouse; cold meats, suckling pig.

   Boar's head, wild duck, boned turkey.

   Quails stuffed with foie gras, medallion of game.

   Asparagus, baked beans, salads.
5. Indian food: Kababs--seekh, shami; tandoori preparations; pulao rice. Vegetarian: dahi-bhalla, potato wada, rice, a few dahi-bhalla,, peas pair, palak panir.

4

Cooking for Patients

> Those who have finished by making all others think with them, have usually been those who began by daring to think with themselves.
>
> —*Charles C. Colton*

Cooking for patients is not a special method of cookery but we have to use our knowledge of baking, boiling, steaming, etc., and in addition follow some rules which apply specially to food prepared for the sick. An invalid usually has a weak appetite and needs tempting because strength can only be regained through taking suitable food. Wrong food may kill a patient, while badly prepared food will retard recovery.

Good food, well cooked and served will do much towards saving the patient's life. Convalescents vary in appetite according to the nature of their complaint. But whether they have good appetites or not, their food should be easily digestible as they cannot take as much exercise as a normal person.

Guidelines

(a) Consult the doctor about the patient's diet and obey his instructions strictly.

(b) Serve food which can be easily digested because in sickness the digestive system is often impaired.

(c) Serve well-balanced meats, but pay special attention

to the constituents according to the nature of the illness. In some cases meat, starches and sugar may have to be taken sparingly or even eliminated.

(d) Choose food that will supply necessary nourishment and be suitable to the illness.

(e) In a serious illness, food is usually given in liquid form and should be as varied as possible.

(f) In convalescence see that the solid food is light, easily digestible and served in small quantities.

(g) Vary the food and the method of cooking as much as possible. This stimulates appetite.

(h) As convalescence proceeds, increase the food in amount and variety.

(i) Serve the best quality possible. Food should be absolutely fresh.

Suggested Dishes

Liquid: Fresh fruit juice, barley water, beef tea, egg flip, lemonade or orangeade.

Soft food: Made with milk, e.g., junket or milk jelly, custard puddings (baked or steamed), blancmange, broths, light soups, rice conjee, etc.

Solid Food: White fish, fowl, rabbit, sweet bread or pancreas, liver dishes (on the advice of a doctor), etc.

Storage Methods

The fundamental principle in the storage of food is to keep it clean, cool and covered. Periodical inspection is a must, and there should be a regular turnover of goods. Daily cleaning, weekly cleaning and spring-cleaning are necessary to keep food free from dust and dirt. Access to food stores must be restricted. No food should be kept in the kitchen or larder, apart from what is required during the day. The storage room should be dry, well-lit, ventilated, vermin-proof and clean.

Cupboards and shelves should be of simple design without dust-collecting ornamentation, unnecessary ledges or panels. Doors should be so designed as to enable the whole shelf to be clearly visible and accessible when doors are opened. Shelves must be constructed for specific purposes. They should be so fixed that they could be easily taken down and re-erected to facilitate regular and thorough cleaning. They should be narrow enough to enable all the goods to be easily accessible. Wide wooden shelves should not be made of planks which are joined, tongued or grooved but of 5 cm. x 1.25 cm. (2" x ½") slats laid with a clearance of 3.75 cm. to 5 cm. (1½"to 2") between each slat. The one nearest to the wall should have a return board to form a lip. The back of the lip should be at least 5 cm. (2") from the wall, to leave room for the cleaning of the wall. Corners must be rounded, particularly where shelves on two walls meet at a corner, to avoid an uncleanable dust trap. As far as possible avoid having shelves above windows.

Special shelving covered with marble, hard stone or tiles should be set aside for such materials as fat. One part of the shelf should be set aside for glass receptacles. No foodstuff, even when in bins, boxes or cartons, should be stored directly on the floor. The bottom shelving should be at least 75 cm (26") above the floor. Containers for food should be provided, the nature of the foodstuff determining the nature of the receptacle required. These should be smooth, impervious, easily cleanable, resistant to wear denting, buckling, pitting, chipping, cracking and should withstand penetration by vermin, and the corrosive action of foods and cleaning compounds. Stainless steel, model metal, vitrified china, glass or plastic should be used for moist unpackaged food or beverages, and galvanized steel for dry packaged and unpackaged foods. Solder must be nontoxic, and welded areas must be as resistant to corrosion as the parent metal. Food containers should stand clear of walls, should be kept scrupulously clean and unless the articles they contain are naturally wet, should be completely dry, both inside and outside.

Segregate food to be used promptly from the food that is to be held for some time. Use food in the order of priority--first in, first out.

Stocks should be inspected once a week. A stock list should be made to tick off inspection dates and to indicate conditions. Keeping food clean not only means keeping it free from visible dirt, but also keeping it in such a way as to prevent the multiplication of germs.

High temperature kills them, and freezing preserves bacteria in suspended animation. Hence food should be kept cool and dry. Perishables should be kept under refrigeration or, when no refrigeration is available, in the larder in closed containers or adequately protected with muslin or wire covers. Windows and doors should have fly-proof meshing.

Guide to Storage of Certain Fruits and Vegetables (R.H. 85 to 90%)

Commodity	*Temperature* °C	*Approximate Storage life (weeks)*	*Approximate post-storage life (days)"*
Apples	2°-3°	18	9
Apricots	-1°-0°	1-2	--
Bananas	11°-13°	3	1
Beans	0°-2°	2-3	--
Beetroot	0°-2°	6-8	--
Brinjal	8°-10°	3-4	--
Cashew apples	0°-2°	5	--
Cherries	-1°-0°	2-3	--
Carjots	0°-2°	13-17	--
Cabbage	0°-2°	9-13	--

Commodity	*Temperature* °C	*Approximate* *Storage life* *(weeks)*	*Approximate* *post - storage* *life (days)"*
Cauliflower	l°-2°	4-6	--
Guavas	8°-10°	4	--
Grapes	0°-2°	8	--
Grapefruit	5.5°-7°	8	3
Gooseberry (Indian)	0°-2°	8	3
Jackfruit	11°-13°	6	--
Lemons	5.5°-7°	6	3
Limes	8°-10°	8	--
Litchi	0°-2°	11	--
Mangoes:			
Alphonso	8°-10°	4	--
Badshapasand,			
Dasharj			
Dilpasand			
Khuddus			
Neelum			
Raspuri Safeda,			
Bangalore	5.5°-7°	4-6	3-4
Mangosteen	4°-5.5°	7	--
Oranges	5.5°-7°	18	--
Onions (R.H.70 to 75%)	0°-2°	17-26	--
Papaya	4°-5.5°	5	3
Passion fruit (purple)	5.5°-7°	3	--
Pears	-1°-0°	12	--
Persimmons	0°-2°	7	--
Peaches	-1°-0°	4-6	--

Commodity	*Temperature* °C	*Approximate Storage life (weeks)*	*Approximate post - storage life (days)"*
Peas	0°-2°	2-3	--
Pineapples	8°-10°	6	--
Plums	-1°-0°	2-4	--
Pomegranates	0°-2°	7-11	1
Pommelos	5.5°-7°	23	--
Potatoes	3°	26-35	--
Radishes	0°-2°	6-8	--
Sapota (Chikkoo)	0°-2°	10	5
Tomatoes			
(a) ripe	4.5°-7°	1-1½	--
(b) mature (turning to yellow)	13°-15.5°	3-4	--
Turnips	0°-2°	2-3	--

* at a temperature of 21°-34°C

Poultry: Dressed and ready to cook poultry deteriorates rapidly unless carefully handled and stored under properly controlled conditions. Freshly killed poultry, after dressing, should be drawn quickly, and wrapped loosely in waxed paper or aluminium foil and refrigerated.

Such poultry can be kept for 2 to 3 days in a refrigerator at 3°C (about 38°F) or less. Frozen poultry should be stored at -18°C (about 0°F) or less, and cooked soon after it has been defrosted. Generally speaking, frozen poultry should be used within 6 months after it has been processed to obtain satisfactory results.

Eggs: Eggs are highly perishable products and care must be taken in handling and storage. Eggs should always be stored in a refrigerator. They may lose as much of their fresh quality in 3 days at room temperature as in 2 weeks in the refrigerator. Store with the large end up. At a range of temperatures and relative humidity varying from -2°C (about 28°F) and 85 per cent R.H. and 2°C (about 35° F) and 75 per cent R.H., eggs may be stored for as long as 9 to 10 months. Frozen eggs should be stored at -18° C (about 0° to 20°F).

Treating with oil or wax before refrigeration also extends shelf life for several months. A pure, colourless, odourless, tasteless mineral oil is used and the processing done under controlled conditions.

Milk: Bottled pasteurized milk can be safely kept for at least 3 to 4 days at refrigerated temperatures preferably between -1° and 4.4° C (about 34° and 40° F).

Know-how of up Keep: Knowing how to store food properly, and for how long you can keep it is very important: the food will be safe to eat; it will retain its flavour and texture; a high level of nutrients will be maintained; and you will not waste money on spoiled items.

Safety First: The three types of bacteria responsible for most cases of food poisoning (Salmonella, Staphylococcus and Streptococcus) are present around us all the time. Normally, they are not trouble some, but if given the right (actually wrong) conditions, bacteria can multiply in a matter of hours to dangerous levels. The contributing factors to bacterial growth-temperatures between 1-52°C (about 50°-125°F) and time To thwart growth, do not be casual about handling food; always work with well-scrubbed utensils, cutting boards—and hands; cook or serve food as soon as possible after removing it from storage; refrigerate food immediately after the meal is over; always take precautionary care. It is often not possible to tell by taste or smell if food is contaminated.

Foods that Need Special Care: Give special attention to preparing and storming poultry, fish and sea food, meat, creamed mixtures, mayonnaise puddings and stuffings. Bacteria find these foods a good media to grow in. In particular: do not stuff poultry the night before; the cold stuffing may not heat up to a safe temperature when you cook the bird Never refrigerate cooked poultry with the stuffing in-store it separately If you have bought cracked eggs, use them only where they will be thoroughly cooked (baked or hard-boiled); Salmonella, which may be present on shells, could contaminate cracked eggs.

Know-how of Frozen Foods: It is safe to refreeze virtually all partially thawed foods if they still have ice crystals in them and are still firm in the centre. However, many foods (ice cream and uncooked baked goods for example) will not maintain top quality. Meat, fish and poultry purposely thawed in the refrigerator and kept no more than one day may be refrozen. However, do not refreeze combination dishes-pies stews etc., that have been thawed. With the exception of fruit and juice concentrates, foods thawed accidentally in the freezer over a period of days (because of a power failure, etc.) should not be refrozen, unless of course, they still have ice crystals. If food is completely thawed (on purpose or by accident), warmed to room temperature and left for more than two hours, throw it out. The exceptions: fruit and juice concentrates, which ferment when spoiled. Throw away fruit if flavour is "off"

Pantry-shelf Storage Temperature: Store food in well ventilated cabinets--not over a cooking range or near refrigerator's exhaust. Use cool areas for storing large amounts of potatoes, onions, etc., and for long-term storage of canned foods.

Time: Though most staples and canned foods will keep indefinitely, buy no more than you expect to use in the recommended storage times given later. While foods will be safe beyond the recommended storage times, flavours will fade and textures will wilt. Date foods. Then check cabinets every six months and use up the oldest items.

Buying: Purchase the freshest-looking packages--messy or shop worn labels indicate old stock. Don't buy cans with swollen ends--the food has gone bad. Dented cans may be purchased, provided they are not punctured.

Home Canning: Use first-quality foods and the best techniques.

Temperature: From 1° to 4.5°C (about 34° to 40°F) is best. Above 4.5°C (40°F), food spoils rapidly. Check temperature with a refrigerator thermometer or an outdoor thermometer.

Time: Use foods quickly--do not depend on maximum storage time.

Wraps: Use foil, plastic wrap or bags, airtight containers. When meat, poultry or fish is bought in a plastic wrapped package, loosen ends of package to dry surface moisture--bacteria grow faster on moist surfaces.

General Care: Clean refrigerator regularly to cut down food odours. Remove spoiled foods immediately so that decay will not be passed on to other food.

Freezer Storage

Temperature: -18°C (about 0°F) or below is best. The maximum temperature should be -15°C (about 5° F). Check temperature with freezer thermometer or outdoor thermometer, or use this rule of thumb: if the freezer does not keep ice cream brick solid, temperature is above the recommended level. If this is the case, do not plan to store food for more than a week or two.

Time: Date foods with an "expiry date" according to maximum storage time recommended below. Longer storage is not dangerous, but flavours and textures begin to deteriorate.

Home-Frozen Foods: Use good-quality foods and proper techniques. Freeze foods quickly in coldest part of freezer, then store in another freezer area.

Commercially Frozen Foods: Do not buy them if they are battered, a sign that they may have been partially thawed. If foods have been partially thawed, then refrozen, use them within a few days.

Wraps: Use foil, plastic bags and wraps, freezer wrap or freezer containers.

5

Sea Foods

> **Do not be selfish. If you have something you do not want, and know someone who has no use for it, give it to that person. In this way you can be generous without expenditure of self-denial and also help another to be the same.**
>
> ***—Elbert Hubbard***

Fish provides many attractive and satisfying dishes. Fish is a high protein food, supplying on an average a little more than 5 gms. of protein per edible ounce. Fish could be got from lakes, rivers or seas. Freshwater fish are considered to be superior in flavour to salt water fish. A fish that has fat is superior in flavour to lean fish. The season also has a bearing on the fish for better taste.

The fat content helps to determine the method of preparation. Fish that has fat will produce superior eating qualities, if baked, or boiled because of their natural fat which keeps them from drying during cooking. Lean fish are best poached, boiled or steamed, the flesh is firm and holds together during the cooking period. Both lean and fish with fat can be sauteed or fried which give excellent results.

Selection of Fish: **While purchasing fresh whole fish, the following points should be observed:**

(1) Quality in fresh fish is determined by the condition of the skin which should be shiny and of good colour.

(2) In white fish, the flesh should be really white, not faintly yellow.

(3) The body should be rather heavy in relation to the length, and the flesh plump and springy.

(4) Eyes should be bright and not sunken.

(5) Gills should be red.

(6) For freshness, the fish should have a pleasant, slightly salty smell, while stale fish has an ammoniac odour, which increases with its staleness.

(7) If the imprint of the fingers remain, after it has been pressed, then it is not fresh.

(8) If the scales come out themselves, then the fish is stale.

(9) Raw flesh should not come away from the bones, as it is the sign of fish deterioration.

(10) The tail should be stiff.

Shellfish: The food value of shellfish is similar to that of fish, as it contains protein, calcium, magnesium, iron, phosphorus and iodine. Whilst shellfish is difficult to digest, it is a good body-building food. Because they are scavengers, shellfish are regarded as a dangerous food and a source of food poisoning. But this is largely due to poor buying and storage, or that they are misused during the cooking process. Debeard is to remove beard from shellfish such as oysters and scallops.

Storage: Where possible, shellfish should be purchased alive, so as to ensure absolute freshness and stored in a cold room, in a box covered with crushed ice and a damp sack and cooked as soon as possible after purchasing.

Shrimps (Crevettes Fr.)

There are several kinds and the best known are the brown and red shrimps. They are available all the year round in the coastal areas. They should have a fresh smell and not be sticky or slimy, which would indicate decay. They are mainly used for decorating fish dishes, making fish sauces, seafood cocktails, hors d'oeuvres, potted shrimps, omelettes, savouries, curries and pakodas.

Crayfish (Ecrevisse Fr.)

A small freshwater lobster found in rivers. Average market length in 3" to 4". Of a dull green or brownish colour above and paler brown or yellowish below. They dig themselves into river banks by day and come out at night for food. Generally used for garnishing cold dishes for buffet work but any recipe given for lobster could be applied.

Prawns (Crevettes Fr.)

There are two principal types, the sword prawn (Crevette rose) which is pink, and the common prawn (Crevette gris) which is pinkish grey and smaller. They are larger and more delicate in flavour than shrimps. Obtainable all the year round, usually cooked, with the same points of quality as for the shrimps. They are used in fish dishes, seafood cocktails, canape Moscovite, salad, hors d'oeuvres and curries.

Scampi (Langoustine Fr.)

Mainly a Mediterranean shellfish. The variety used here is the "Dublin Bay Prawn'. They resemble a small lobster, about 8" long and only the tail part is used. Usually purchased in the frozen state as it is difficult to obtain fresh. Main uses are fried, e.g. meuniers, a 1'Anglaise, a la Franchise, a 1'Orly, or poached and finished, e.g. Newburg.

Lobster (Howard Fr.)

Probably the most highly rated of all shellfish. Is of rich dark blue colour tinged with purple. Grows very slowly and when 5 years old will measure some five inches overall, having shed its tail some 20 times in the course of this period. Specimens of 12 to 14 inches long of 5 kg. have been known. They should be alive when purchased and should have both the claws attached. Weight in relation to size is extremely important when purchasing. Parasites like barnacles clustered on the shell indicate age. The flesh of the cock lobster is thought superior to the flesh of the hen, Hen lobsters have a broader tail than cock lobsters. Lobsters are served cold in cocktails, hors d'oeuvres, salads, sandwiches, and on buffets. Hot preparations include soup, grilled and served in numerous dishes with various sauces and also as garnish to fish dishes.

Crawfish (Langouste Fr.)

Greenish brown with yellow markings on tail. This turns to brick red after cooking. They have no claws but have long antennae. Owing to their size and appearance, they are used mainly on cold buffets, but hot lobster recipes can be applied, points of quality and season would be the same as for lobster.

Scallop or Scollop (Coquille de St. Jacques Fr.)

A species of bivalve molluscs with ribbed rounded shells. When fresh, the shells are tightly closed and they should feel heavy for their size. Whilst they are normally very muddy, they should not smell unpleasant nor be covered with barnacles, which indicate age. Method of cooking are usually poaching or frying, dressed back in the cleaned top half of the shell with a border of Duchesse potato piped around and finished with a sauce and garnish.

Mussels (Monies Fr.)

An edible mollusc, of which there are two main species, the common mussel from the cold regions and the large mussel from the Mediterranean. A distinction must also be made between wild mussels gathered in their natural beds and rocks, and cultivated mussels, which are tender and delicately flavoured, small but very plump. Mussels can cause skin eruptions and digestive disturbances. However, as they are always cooked, dangerous microbes are destroyed.

When purchasing, shells should be tightly closed, large, and not having an excessive number of barnacles and no offensive smell. After purchase, they are best kept in a box of crushed ice with a damp sack, and placed in a cold room. Maybe served raw, hot, or cold, as hors d'oeuvres, cocktail or garnish.

Crab (Crabe Fr.)

A shellfish of many varieties, fresh and salt water. Common varieties known as the edible crab. They should be alive when purchased and both the claws attached to the body. The claws should be large and heavy. There is usually more flesh on the hen crab, but it is considered of inferior quality to the cock. Used for hors d'oeuvres, cocktails, salads, dressed crab sandwiches and bounchees.

Cockles

These little shellfish are marine molluscs found mainly in the mouth of rivers. The white fish is enclosed in two concave shells of equal size, white in colour. Cockles are traditional delicacies.

They are raked up from sandbanks when the tide is out, and they are usually sold boiled and shelled. The flavour is not as delicate as other shellfish, but is interesting and savoury. The type found on the Atlantic coast of France, called Bucarde-Sourdone, is edible. It is also known in France under the name of conque, sourdon and poor man's oyster.

Cockles are eaten raw and can also be prepared like mussels.

Winkle or Periwinkle (Bigorneau)

A name applied to many small gastropod molluscs. Winkle farms have been set up where winkle are cultivated.

Generally speaking winkles are eaten raw, but they can be cooked like cockles. To eat it, the winkle is extracted from its shell with a pin.

Clam

A name applied to many edible bivalve molluscs (hard-shelled animal with two halves) which live in slimy sand and are found at low tide. Some of those clams are cultivated in special beds along beaches. Auray and Roscoff are the principal French centres of the clam trade.

Clams are mainly eaten raw but they can be cooked in the same manner as mussels, oysters and many other shellfish. The North American people are very fond of this shellfish, and it is prepared in innumerable ways.

There are two types of clam--soft-shelled clams and hard-shelled clams.

Oyster

The oyster has been known to man from the earliest times. Nevertheless, there is no positive evidence that the Egyptians and Assyrians ever ate this delicious mollusc. It seems to have been unknown to the Jews, also there was no mention of them in the Bible. On the other hand, the Celts gathered oysters and fed on them abundantly.

In France, right up to the nineteenth century, it was thought that the coastal oyster-beds were inexhaustible. So much so that the Ordinance of 1681, which was enacted to protect mussels, left the oyster fisheries free from control. The natural oyster-beds continued in all coastal areas and it became necessary to guard the oyster-beds in the bay.

Nutritive Value: Raw oysters contain vitamins,

phosphorus salts, chalk, iron, copper and manganese in a readily assimilable form. They also contain a very high proportion of iodine and are sometimes eaten to combat anaemia. The risks of infectious diseases being spread through the water contained in oysters is small. Tradition has established the hygienic rule that oysters may not be eaten during the warm months of the year--that is to say, when there is no 'R' in the name of the month.

Oysters are "in season" from September to May, although the idea that they are inedible during the summer months, the breeding period, is erroneous. Pacific coast oysters are sold all year long.

Types of Oysters: There are two types, the wild oyster, which has a very pronounced sea water flavour and the cultivated oyster. In the cultivation of oysters, seed-oysters are affixed to tiles and reared in farms some way out to sea. When they have reached a certain size, they are transferred to the fattening beds, which are always situated at the mouth of a river, the mixture of fresh and sea water being essential to induce overgrowth of the liver in which the fattening consists. Whitsable, Colchester and Helford are the chief British centres of oyster cultivation.

Storage: Oysters are stored in barrels or boxes, covered with crushed ice or damp sacks and kept in a cold room.

Fish should be stored for a long duration at -18°C to -24°C.

Quality Determination

1. Oysters must be alive; this is indicated by the firmly closed shells.
2. They are graded in size and the price varies accordingly.
3. Oysters should smell fresh.

4. They should be purchased daily.
5. They are in season from September to May.

Indian Varieties: The production of fish in India as compared to other countries is rather low. The major fish of India are sardines, mackerel, prawns, tuna, etc.

Pre-preparation

Descaling and Cleaning

(1) Soaking the fish in cold water for a few minutes before descaling, helps in removing scales more easily.
(2) The blunt side of the knife should be used.
(3) The head of the fish is held with the left hand and holding the knife vertical, scraping is done starting from the tail, working towards the head, the scales are scraped off. The fish is then washed to remove any loose scales.
(4) Cut off the fins, remove the head. The entrails should be removed by cutting the length of the fish from the vent end to the head on the belly side.

Filleting: Cut the flesh along the line of the backbone and raise the fillet from the middle of the back, to the sides, first working towards the head, then the tall.

Skinning: Hold the tail end of the fish in the left hand, first sprinkling salt on the fingers for a good grip. Skin the flesh (skin side down) from tail to head, with quick short sawing strokes of a sharp knife. Point the knife blade towards the skin so that no flesh is wasted.

Cuts of Fish: The common cuts of fish are:

Fillets	Deboned, long fiat pieces of fish, without the skin. Flat fish gives 4 fillets.
Paupiette	The fillet of fish is laid flat on the table, stuffing is

	spread on it, then it is rolled up, tied with a string, to keep the shape. Can be poached.
Supreme	Large fillets of fish cut on a slant of a large round or flat fish e.g. supreme of turbot.
Meuniere	Pan frying of fish, finished with butter noisette, chopped parsley and lemon juice.
Goujons and Goujonettes	Strips approx. 8 cm long by 1 cm. wide cut from fillets of fish. Goujonettes are cut smaller and are mainly used as a garnish. Can be crumbed and deep fried.
Tron, con	It is thick piece of fish 4-5 cm. thick, cut on the bone from a flat fish like the turbot. It is a fish cutlet or steak with bone.
Darne	A piece of fish cut across and through the bone of a large, whole round fish such as cod, salmon which is 2-3 cm. thick.
Mignon	Fillet of fish folded as a cornet (triangular fold as for piping bag).
Phe	Fillet flattened and folded in two.
Medallion	A medallion shaped portion from large fillets.
Fish Farce	Puree of fish, bound with whole eggs panada and cream.

Significant Liquors Used

Court Bouillon: To preserve and enhance the delicate flavours of fish, the liquors in which they are poached are prepared with care. The poaching liquor is called court bouillon. Different types of court bouillon are:

White Court Bouillon: Sliced onion, bay leaf, parsley stalk, peppercorns, juice of lemon, salt, milk and water.

Ordinary Court Bouillon: Ingredients--Vinegar, water, peppercorns, sliced carrots, onions, salt, parsley stalks, thyme and bay leaf.

White Wine Court Bouillon: Ingredients--White wine, water, minced onions, bouquet garni, salt, peppercorns, sliced carrots.

Red Wine Court Bouillon: Ingredients--Red wine, water, minced onions, bouquet garni, parsley stalks, salt, peppercorns and sliced carrots.

Method: For all the different court bouillons, the ingredients are simmered, strained and used for poaching, boiling, stewing, etc.

Special Application

The method of cooking depends upon the characteristics of the fish, its texture, size, form, percentage of fat and its strength of flavour. Over cooking of fish should be avoided. Oily fish are best for baking and grilling and lean fish are good for steaming, poaching, etc. Either lean or fish with fat are suitable for all types of frying,

Boiling: The method is suitable for the whole fish, e.g. Malli, Surmai, Indian Salmon and Trouts, etc. Thick cuts of fish could also be boiled. Whole fish cleaned could be put in court bouillon and boiled. The other method is to wrap the whole fish with cheese cloth (to prevent it from breaking), and to retain the shape, could be tied to a wooden plank. This is put into a pan, large enough for the fish and court bouillon used, as the cooking liquor. It should be sufficient to cover the fish and should be brought to the boil and then the temperature should be lowered and allowed to simmer. If the fish is cooked, then a skewer will easily pierce. 10-15 minutes per 500 gms of fish is enough to cook the fish.

Bisque: It is a rich classical soup made from shellfish, shells, tomato, brandy and cream.

Steaming: It is necessary for steaming, the fish should be well seasoned and sprinkled with lemon juice to keep the flesh white. Greased, grease-proof paper should be put at the bottom of the steamer before cooking. This helps in lifting the flesh out and does not allow it to stick to the steamer, before cooking. The fish does not stick or break, and can be easily lifted. Cooking time varies from 10 minutes for fillets to 20 minutes for cutlets, and larger pieces. Steamed fish must be well drained before service and should be accompanied by a good, sharp, well seasoned sauce.

Poaching: This method of cooking is suitable for small whole fish, cuts or fillets. Court bouillon should be used for poaching as it enhances the flavour of the fish. Cooking liquor barely covers the fish, covered with grease-proof paper and poached in the oven. In a moderate oven, it takes 15-20 minutes and the cooking liquor could be used for making sauces.

Baking: Round white fish is suitable for baking. The fish should be cleaned, and the belly is stuffed with force meat (breadcrumbs, chopped onions, nutmeg, lemon rind, egg yolk, a little butter, seasoning) or any other stuffing and the edges are sewed. Butter is applied to the fish and then it is baked in a baking tray. Fish augratin and fish souffle are also baked. Baking time is 10 minutes per 500 gms.

Frying: Small whole round of fiat fish, goujons and fillets are suitable for deep frying. As the flesh is very delicate, the fish must be coated before frying. The following are the coatings normally used:

(1) Flour, egg white and breadcrumbs
(2) Milk and flour
(3) Flour batter
(4) Flour and beaten egg

Fillets of fish are always coated with flour and then dipped into any of the above protective coating and the flour helps the coat to stick. The cooking time is about 5 minutes.

Shallow Frying: Fillets, tron, cons, supremes, steaks

or small fish are shallow fried. The term 'meuniere' is used when fish is shallow fried. The fish is coated with seasoned flour and shallow fried in butter. The cooked fish is masked with melted brown butter and garnished with fried or fresh parsley and lemon slices.

Grilling: It is a suitable method for various cuts and types of fish. They are seasoned, oil applied and cooked over heat on the grill or in a salamander. The grill should be greased, so that the fish does not stick. Only fish do not need extra basting. Grilled fish is garnished with lemon and parsley and accompanied with a suitable sauce or butter. The time taken or grilling is about 5 to 7 minutes.

Stewing: Bouillabaise and Matelote are French regional stews in which a mixture of different fish are used. Pieces of different fish and shellfish with vegetables, herbs and fish stock are cooked and the cooking liquid is not thickened.

Fish Kettle: Long narrow, deep pan used for poaching.

Different Fish Dishes

Pomfret

Pakodas: Dip pieces in garlic flavoured besan batter and deep fry, serve with pudina (mint) chutney.

Diable: Spread with mustard, sprinkle with breadcrumbs and grill. Serve with Sauce Ravigote.

Maitre d'Hotel: Serve parsley butter *(Maitre d'hotel Fr.)* with fried, poached or grille fillets.

Flamande: Fillets poached in white wine, mushrooms, parsley, chopped shallots Reduce the sauce, garnish with slices of lemon.

Portugaise: Poach in white wine, chopped onions, garlic, parsley and tomato concasse Dish and coat with reduced sauce.

Trout (Fr. Truite)

Bretonne: Cook the fillets in butter and serve with finely minced mushrooms and salted shrimps.

Frite: Paner al'Anglaise, small fillets, deep fry and garnish with lemon slices.

Grenobloise: Coat the fillets in seasoned flour, cook in butter, pour brown melte butter (beurre noisette Fr.) garnish with lemon slices and capers.

Meuniere: Coat the fillets in seasoned flour, cook in butter, garnish with lemon slices chopped parsley and melted brown butter poured on top.

Aigulette Murat: Fillets cut into strips coated with seasoned flour and sauteed in butter, served with strips of artichoke bottoms and Parmentiere potatoes.

Tandoori: Marinade in yoghurt, garlic, ginger, green chilli paste, orange colour, garam masala and salt. Skewer the fish and place in tandoor (mud coated oven and wood /coal is the fuel used) and roast it. Apply hot butter and serve with onion rings and lemon slices.

Indian Salmon

Coquille Mornay: Poached and placed in shells, coated with mornay sauce and bordered with duchesse potatoes and gratinated.

Espagnole: Fillets cooked in butter placed on sauteed tomato concasse and served with fried onions and sauteed juliennes of capsicum.

Ouglerce: Fillets poached in white wine, fish fumet, chopped onions, parsley and tomato concuss and remove the fillets, reduce the sauce, cook and mix with veloute sauce, coat the fillets with the sauce and serve cold.

Bercy: Poach the fillets with chopped onions, parsley,

white wine and fish stock. Reduce, add butter and flour *(Beurre manie Fr.)* cook, coat the fish and gratinate.

Veronique: Poach fillets in fish stock, reduce, add butter, garnish with destoned muscat grapes coated with sauce and gratinated.

Nori: Glossy purple black seaweed used in sushi and cold fish dishes.

Amritsari Machchi (Fish)

Fish covered with masala, i.e., amchoor (dried mango powder), besan, red chillies powder, jeera, ginger and garlic seasonings. The average serve of whole fish on a table d'hote menu is 200-250 gm.

6

Dishes of Eggs

The most popular and widely used eggs are of hens in India, although eggs of turkeys, guinea fowls, ducks and geese are also used. Eggs have been described as a "Complete food" because they are highly nutritious and used in a range of recipes, yolks and eggs are used separately.

The colour of egg shells and egg yolks varies with the breed of hen and the food they eat; but this makes no difference to the food value. They are rich in proteins and contain valuable amounts of iron and vitamins A & B. Eggs are nourishing and easily digestible. As such, they are given to invalids and infants. The ideal storage temperature for eggs is 2°C-5°C.

Eggs are essential to all kinds of cooking, not only in the preparation but as a food in their awn right. They produce meals that are economical in price, and with a minimum of waste and time. By the term eggs, we mean those of the domestic hen, but these are not the only edible eggs. There are those of the ducks, geese, turkeys, also of wild birds such as plovers and gulls. Other eggs eaten as food are a variety of fish, i.e. (roe) herrings and sturgeon (caviars).

A Valuable Food: An egg is a compact and almost complete food. It is equal to seventy calories of energy and contains carbohydrates, fats, proteins, mineral salts, vitamins and water. Vitamins A, B, D are all necessary for health and growth. Iron, phosphorus and calcium are essential for building and maintaining healthy bodies. These furnish all the essential amino acids for building and repairing body tissues. From the culinary point of view, an egg comprises two separate parts. The white or albumen and the yolk. The average egg is 10% shell, 30% yolk, 60% white; 73% water is contained in its constituents.

To test the freshness, place it in water. A bad egg will float, a good one will sink and varying degrees of freshness would be indicated between these two positions.

Uses: In all culinary preparations eggs are used for binding, colouring, enriching, increasing volume, coating, etc.

Strengthening an Egg White: This is done by adding a pinch of salt/Corn-flour/or arrowroot to the whites just before reaching their peak to stiffen the final result.

Storing: The shell of an egg is very porous and for this reason, eggs must be stored away from contaminating odours. Keep in a cool dry place. Ideal temperature: approx. 2°C-5°C. Fresh eggs may be preserved and stored in water glass, a gelatinous odourless substance.

Remember, the dirty grey rings which appear when hard-boiled eggs are sliced, are caused by the iron in the yolk and the sulphur in the white combining under certain conditions to form a black chemical compound called iron sulphide. This can easily be overcome j if the right methods are followed.

Separating Whites and Yolk: If a little of the yolk gets into the white when separating, remove with a piece of the broken shell. If the whites are for whipping this is important, as the white will not whip, the slightest smear of grease on any of the utensils will also prevent the white from whipping to a stiff froth. Use only copper, stainless steel, or china bowls for whisking egg whites.

Cooking of Eggs: The basic principle of egg cooking centres around the ability of the proteins in both the yolk and the white to coagulate when heated. The rate of coagulation 1 and the firmness of the gel depends on the intensity of heat, cooking period and the presence of other Ingredients such as milk, water, sugar and salt. Sugar increases the temperature at which coagulation takes place, and acid and salt speed up the process of coagulation. Extended cooking or high heat toughens eggs. While boiling eggs, if high temperature and long cooking is applied, the iron and sulphur compounds in the white are released to form ferrous sulphide. It is a blackish or purple ring around the yolk. To avoid this, as it spoils the appearance of the eggs, fresh eggs should be used and simmered. When the eggs are boiled, they should be placed in cold water, to release the pressure inside that discolours. The time required for soft-boiled eggs is 3 to 5 minutes and hard-boiled eggs about 10 minutes.

The coagulation quality of eggs makes them usable for various purposes and thus they form an important raw material for cooking. They can also be combined with other ingredients to produce a variety of dishes.

Forms of Using: Various forms of using eggs in food preparation are:

Binding: Addition of eggs to minced meat and mashed vegetables, etc. helps to bind the mixture. As the heat coagulates, the proteins are bound into a cohesive mass. It helps to retain the shape of mutton croquettes, meat loaf, medallions, hamburgers, koftas, kababs, etc.

Coating: The egg or egg batter help to give a coat to the food items and prevent them from disintegrating and give them a protective coating. Many of the food items such as fish fillets, croquettes, cutlets, etc. are dipped into the batter before crumbing and then fried. Eggs are also used for preparing pancake batters (eggs, flour and milk).

Leavening: By beating the egg whites, a foam is made up of air bubbles, surrounded by a thin elastic film of egg white. This mixture, when added to products such as sponge cakes, meringues, souffles, etc. increases the volume and the egg white film hardens. Addition of sugar to egg whites, makes it stable, smooth and the foam does not collapse easily. Egg yolk has a less foaming power because of its fat content. Egg is used as the principle ingredient for chou paste from which eclairs, beignets, fritters and proflteroles are made.

Emulsifying: Eggs form stable emulsions. For example, mayonnaise, oil and vinegar separate out unless oil droplets are coated with substance, that keeps them from separating. Eggs are the emulsifiers that give a smooth mayonnaise sauce. It is also used as an emulsifier in icecreams, cakes, cream puffs.

Eggs enhance colour and sheen. An egg wash gives a shiny appearance to pastry and baked dough products. Eggs also enrich and enhance the flavour and taste.

Cakes and farinaceous dishes are enriched by adding eggs, which improve the flavour and taste.

Thickening: Eggs help to improve the consistency of gravies, curries, sauces and soups. Egg liaisons used in soups and sauces help to thicken and Improve the consistency. When used in custards, the heat coagulates the eggs and makes the custard firm.

Garnishing and Decoration: Slices, sieved or quarters of boiled eggs are used to decorate or garnish dishes such as--salads, biryanis, curries, Vienna steaks, etc. For Consomme Xavier, threaded eggs are added as a garnish, and for prawn cocktails eggs are the garnish.

Various Courses in which Eggs are used:

Hors d'oeuvre: Used for various hors d'oeuvres, such as egg mayonnaise, egg stuffed, egg ducklings, etc.

Soups: In the preparation of consommes, thickening of veloutes, for garnishes such as slices of savoury custard (royal), threaded eggs (Xaviere).

Eggs: They are placed as the third course before the fish. They are featured only on luncheon menus and not dinner. The egg dishes served are scrambled, Encocotte, Egg Portugaise, etc.

Farinaceous: For raviolis, cannellonies, and noodles, eggs are used to prepare the paste.

Fish: Batters are prepared out of eggs for the coating prior to crumbing.

Meat and Poultry: Chicken cutlets, hamburgers, meat, beef, Vienna steaks, get their binding from eggs.

Salads: Hard-boiled eggs are used in salads as main ingredient, border or as a garnish.

Sweets: Eggs bind and enrich the pudding and sweet dishes.

Savouries: Scotch woodcock, Welsh rare bit, cheese souffle, devilled eggs, are all made with eggs.

Cooking Method

Boiling: Lower the egg in boiling water without cracking shells-- four minutes for soft boiled eggs and ten to fifteen minutes for hard boiled eggs. Cool quickly if it is to be served cold. It prevents green rim appearing around the set yolk and stops egg cooking.

Poaching: The egg is cooked in simmering water to which a little salt or vinegar has been added to help set the yolk more quickly. Break the egg and lower carefully in the water and use an egg ring to keep shape.

Scrambling: Beaten egg and milk is cooked over a low heat in a little melted butter until set, but not too firm. Stir the mixture constantly to distribute the white and yolk throughout. Add salt after cooking to prevent drying and toughening of the protein.

Frying: Preheat the pan to a moderate temperature and cook the egg gently to avoid the bottom browning or catching while the yolk remain unset. Never add eggs to a hot pan or the whites will cook quickly.

Omelette: Stir mixture to facilitate even cooking so that the eggs in centre and edge cook evenly.

Baking: Place egg in a ramekin dish, cover with cream to prevent drying. Cook in moderate oven for 6 to 8 mins.

7
Cereals and Pulses

Rice and pasta feature in many cuisines as either a staple food or because of their bland taste, as a traditional accompaniment to spicy or strongly flavoured dishes. As both are economical items to either purchase or make, they can be included in various recipes at a reasonable cost, which is helpful to the caterer. As pasta is available in many forms it can be used with some flexibility in a range of sweet and savoury food.

Rice

It is a widely used cereal which is available in the following forms:-

1. *White Rice:* It is a popular variety and is purchased in both grain for desserts and long grain for main dishes. It has a bland and soft texture.
2. *Brown Rice:* The total grain, inclusive of the outer husk which makes it more nutritious alternative to white rice. It is available in both short and long grain and requires a slightly longer cooking time to soften the husk, although it will never achieve the same softness as white rice. It has a nutty taste and slightly chewy on the palate.

Cooking Rice: The objective when cooking rice is to keep the grains separate, thereby producing a light fluffy result. When the rice cooks, it absorbs water and expands, while at the same time releasing starch into the water. As a general rule, rice multiplies three times during the cooking which implies that careful calculation should be used to achieve the appropriate quantity of cooked rice.

Boiling Rice: The objective when cooking is to keep grains separate and getting a light fluffy result. When rice cooks, it absorbs water and expands, while at the same time releasing starch into the water.

Two methods can be used for boiling rice, one method is to throw away the excess water by straining the rice. The second method is the absorption method by using less water, a ratio of two to one.

Different dishes are made of rice, example: rice soup, fried rice, rice pilaff, mexican rice, etc.

Using Rice in Recipes: Rice can also be used in some of the following preparations:-

Puddings: Uncooked rice is placed in a casserole and an egg custard mixture is poured over the top. It is cooked slowly for around one and a half to two hours in a water bath until the custard is set.

Rice Salad: Cool boiled rice is mixed with other ingredients and in a well flavoured dressing. The salad could be made in advance or made with leftovers.

Binding: For meat loaves, meat balls or croquettes, kofta etc, cooked rice can be added to the rice to thicken, bind or to give substance. Rice flour is used to thicken or bind dishes.

Filling for Vegetables: E.g. Stuffed peppers, tomatoes, zucchini (turai) dolmades or as an ingredient for stuffings of roast meat.

Cooking Pasta: Allow plenty of water in a large saucepan

about eight times the amount of pasta being cooked. Pasta tends to double during the cooking.

Add a good pinch of salt to the water along with a little oil, and bring to a rolling boil. Add the pasta and cook for fifteen to twenty minutes until, 'aldente' that is it is firm when bitten through the centre. Lasagna sheets should be popped into the boiling water individually and allowed to cook separately for a minute to prevent sticking. Once the pasta is cooked, it is drained and rinsed in either hot or cold water, depending on its intended use.

Avoid leaving the pasta in a colander or it will tend to stick together on cooling. It should be served immediately.

Pasta can be reheated in the same way as rice, that is by steaming or in the microwave. It will last for one or two days once cooked and can be frozen as part of a prepared recipe for several months.

There are many sauces that are traditionally served with pasta, ranging from the popular Bolognese (rich meat sauce) or marinara (with seafood) to carbonara (with beaten egg, bacon and cream). The variety of sauces used is almost endless, like the pastas available. Sauces are used to escalate the nutritional value of the dish.

Pasta, like rice is a high carbohydrate food and the addition of a protein based sauce consisting of meat, fish, chicken, or eggs will instantly boost its balance of nutrients.

The predominant flavour in many pasta dishes is Italian in influence and garlic, tomatoes and herbs, such as oregano or basil feature regularly.

Catering Considerations: Pasta can be used as a substitute for potato in a recipe and it is a filling dish, like rice. Serve only one course of pasta. Avoid serving large quantities. Pasta can be used as an ingredient in soups, salads, entrees, and casseroles as well as the traditional recipes. Allow 40 gms of uncooked pasta per person for main course 20 gms per person for a first course.

Catering Considerations: When served as an accompaniment to casseroles, curries or other foods containing a sauce it is used to mop up the sauce that would otherwise run across the plate.

When planning menus it is wise to remember that rice can be used as substitute for potatoes, should never be served together in the same meal. Avoid repeating rice in courses. It is ideal in all seasons and can be served as a salad in summer or as a pilaff in winter.

Pasta: It is a versatile food that can be an ingredient in many courses, although it is usually associated with savoury dishes. Home made pasta is easily prepared from a simple dough and if handled correctly, can provide a flavour some alternative to the commercial varieties. Each type of pasta differs in size, shape and sometimes colour. The addition of various ingredients like finely chopped spinach (green) tomato paste (red) and carrot juice (orange) will produce different effects. Own pasta can be made in a pasta machine.

The main ingredient of pasta is flour, which should be a strong variety, usually made from durum wheat. Wholemeal flour is proving a popular alternative to the plain dough although it should be remembered that extra water may be required to produce a soft dough and kneaded lightly. Eggs are sometimes added to a smooth dough. Water and oil are to be added to the flour which is then mixed to a smooth dough and kneaded lightly. Eggs are sometimes added to the basic dough for flavour and colour.

Once the pasta has been prepared, it is rolled out thinly. Cut into the desired shapes and left to dry out before cooking. When it is completely dried out, it can be stored in an airtight container.

Varieties of Pasta

Bows: Squares of paste cut with a crimped edge and pinched in the centre.

Shells: Resembles a seashell with a wide, curved side.

Macaroni: Small thin tubes with a slight curve.

Twists: Two thin strands of pasta that are twisted together.

Spirals: Single, wider strands of pasta that are twisted.

Tagliatellelf Fettuccine: Long, flat, ribbon shapes of varying widths.

Rigatoni: Wider tubes than macaroni and have a ridged edge.

Wheels: Circles that have a spoked pattern through them.

Ravioli: Small squares of pasta that have a savoury filling between two layers of pasta.

Spaghetti: Long thin strands.

Vermicelli: Thin strands that are wound together into nests.

8

Dishes of Meat

In order to know how the animals are slaughtered and to buy meat sold in a sanitary condition, it is important to know how the animals are slaughtered in hygienic conditions. Meat bought must be from a slaughter house that has professionals working.

Slaughtering and Butchery: In Europe slaughter houses are called abattoirs. The slaughter houses have space for meat storage and deal with large amounts of offals. A slaughter house is functional, compact in space, has satisfactory working conditions, proper equipment and has provision for the highest sanitary standards.

The animals are slaughtered by professionals in a systematic way.

Pre-slaughter Steps

Inspection: Animals are inspected by medical inspectors to assess whether they are fit for human consumption; then only, they are deemed fit for slaughtering.

Resting of Animals: The animals are kept in clean dry enclosures (pens) to rest for 24 hours. The animals are given plenty of water during resting,

(a) It helps to maintain normal heartbeat of the animals.
(b) It helps in the drainage of blood.
(c) It helps to have normal body temperature as it

increases the keeping quality of the carcass and minimises the bacterial growth.

(d) It speeds up the glycogen stored in the muscle to turn to lactic acid, which lower the pH of the muscles and reduce the growth of bacteria and act as a natural preservative.

(e) It shortens the rigour mortis period. (The carcass is hung in cool temperature during the period, the muscles become stiff. Certain enzymes/acids are produced that soften the meat.) Technically It is called ripening or ageing. The meat is tenderised because of the action of the acids. In a fatigued animal the rigour mortis period is longer, whereas in a rested animal it is short.

Fasting: Animals are fasted for at least 8 to 12 hours.

(a) It helps to reduce the amount of undigested food and faeces in the intestinal tract.

(b) Bacterial infection are minimised.

(c) It improves the quality of the meat. The blood left in the tissues is free from absorbed nutrients.

Washing: After resting and fasting, the animals are passed through a tunnel containing showers from top and bottom with lukewarm water and the animal is kept on a sliding platform and is washed from all sides and superficial dirt is removed.

Stunning: It is to render the animal unconscious so that it may bleed freely and it may easily be killed. It is also done to avoid pain to the animal, while it is being killed.

Different ways of stunning are:

(a) *Knocking:* The animal is hit with a heavy hammer (4 lbs.) on the head to make it unconscious. Not a very popular method nowadays.

(b) *Carbon Dioxide Chamber:* The animals are made unconscious by putting them in these chambers, but not popular as the weak animals may die.

(c) *Cartridge Pistols/Captive Bolt:* This is done by damaging the brain by placing the pistol between the eyes, this renders the animal unconscious.

(d) *Electric Tongs:* This is the most popular method. The ends of tongs are placed behind the ears and a current (60-70 volts) is passed. This renders the animal unconscious for about 15-20 minutes.

Sticking: Sticking is done with a knife. 6 inches long blade and it is spear pointed. It is done soon after stunning when the rate of the heartbeat and the blood pressure are high. This assists bleeding. In India, Muslims and some other communities take meat from animals slaughtered" by the Halal method in which throat is quickly cut with a sharp knife with short strokes and bled. The jugular veins and carotid arteries are cut. Sikhs only accept meat which is slaughtered by the Jhatka method, i.e. the head is severed from the body by a single blow from a long, heavy, butcher's knife.

Bleeding: The animals are bled, to the fullest possible extent. The animals are hung on hooks with their heads downwards, as neither the animal nor the slaughterer comes in contact with the blood. Drains for the blood lead outside the building either to a blood collecting tank or to a processing room.

Flaying and Cleaning: Air is blown between the flesh and skin to create space and the skin is pulled off. The stomach and intestines are removed, inspected and sent to the tripery section. Offals are removed and sent to the offal section.

The carcass is now split into hindquarters and forequarters or left whole and sent to the butchery section.

Meat: The different types of meat are: Lamb, Mutton, Beef, Veal, Pork and Game meat.

In order to cook meat properly, it is important to know the composition and the structure of meat.

Muscle tissue is composed of very minute fibres bound together in bundles. They are tiny tubes filled with water, in which are held the mineral constituents, vitamins and emulsified fat, protein of meat. The muscle fibres are held by connective tissues which are of two kinds. The white connective tissue is collagen, and the yellow connective tissue is elastin.

Quality Factors

Fat: The fat present in meat indicates the quality, as it gives a good flavour. Fat in animals is found around the music fibres or between muscle fibres. It is interspersed between the muscle fibres. The fat seen as small flecks of fat within the muscle is known as marbling and enhances the flavour of meat. Fat contributes moistness to the meat, makes it tender, enhances flavour and makes it succulent.

Colour of the Meat; Older animals have darker coloured meat and a yellowish fat, whereas the younger animals have a lighter coloured meat with a firm, dry, creamy white fat.

Tenderness: The tenderness of meat depends upon the food the animal is fed on. The finer the grains, the more tender is the meat. The amount of connective tissue is directly related to the tenderness of meat. Cuts of meat with much connective tissue are tougher I than those containing little connective tissue. The least used muscles such as are found in the loin and rib cuts, are more tender than those that are fully developed such as those found in neck, chuck, breast, etc.

Factors Responsible for Tender Meat

Moist Heat: White connective tissue called collagen changes into gelatine when moist heat is applied to It Yellow connective tissue is called elastin, and this does not change by cooking or marinading. It should be either discarded or finely minced or chopped, which breaks it down.

Tenderizers: When used, tenderizers make the meat tender. Commercial tenderizers contain proteolytic enzyme, such as papain, an enzyme found in raw papaya and its leaf. Raw papaya paste applied to meat helps to tenderize it. Acidic foods like vinegar, tomatoes, curds, lemon, tamarind help to tenderize the food also.

Ripening or Ageing of Meat: Meat should not be eaten immediately after killing as rigour mortis sets in and stiffens the muscles. The muscle is hung in cool conditions 1°-2°C (34°-36°), the time period varies depending upon the type of meat, i.e. beef, mutton etc. There is an improvement in tenderness, flavour moisture and colour in the meat. Veal and pork are not hung.

Marinating: Tough meats soaked in an acidic solution, i.e., wine, vinegar, curds make the meat tender and enhance the flavour.

Mechanical Pounding and Grinding: Mechanical pounding and grinding is done to meat to break down the connective tissue. Grinding breaks and cuts the muscle fibres and connective tissue making it possible for all ground meat to be prepared in a fashion similar to that used for tender cuts. Pounding is used to tenderize the meat; this process breaks and tears only the surface meat fibres and connective tissue.

Cooking Methods: The method of cooking which is most suitable for a particular piece of meat is decided by the kind of animal from which it comes, its condition, age, quality, its fat content, the cut of the animals and the size of the cut and its tenderness. The tenderness and the amount of fat in the tissues influence cooking method and eating quality. Developed muscle fibres usually take longer to cook. A suitable cooking method should be used for various cuts of meat, that preserve nutrients, develop tenderness, preserve juiciness, thus resulting in the least shrinkage.

Moist Heat: Braising, boiling or stewing are methods of cooking in which liquid is the medium of cooking. Collagen in the meat, when subjected to moist heat, changes into gelatine. For braising, the meat is browned and then placed in the oven between 250°-300°F for cooking. Medium heat makes the meat tender.

Steaming: Meat cooked in a steamer is less tender and juicy and does not have such a good flavour. The moisture in the tissues of the cuts of meat changes the collagen or connective tissue into gelatine and tenderizes the meat.

Dry Heat Roasting: For roasting, a constant temperature of 250°-300°F should be used, till the internal temperature of the meat has reached 180°F, as it makes meat palatable and enhances flavour and there is less shrinkage. After it is cooked, the meat should be browned by increasing the oven temperature. A meat thermometer helps in determining the degree

Approximate Roasting Time for Meat

Beef - - 15 to 20 minutes per 455 gins, and 20 mins. over.

Mutton - - 25 mins. per 455 gms. and 20 mins. over.

Lamb - - 20 mins. per 455 gms. and 20 mins. over.

Veal - - 25 mins. per 455 gms. and 20 mins. over.

Pork - - 25 mins. per 455 gms. and 25 mins. over.

Internal Temperature Recommended for Roasting Meat

Beef	*Mutton*
Rare 60°C (140°F)	60°C (140°F)
Medium 71°C (160°F)	71°C (160°F)
Well done 82°-85°C	82°-85°C
(180°-185°F)	(180°-185°C)
All pork meat is well done.	

Grilling: Tender cuts of meat are used for grilling such as cutlets, boti kababs, seekh kababs, chops, beef, ham steaks and bacon. The temperature and distance of the grid on-which meat is cooked should be varied according to the thickness of the meat. A thin piece can be grilled by turning only once and thick pieces, a number of times. Oiling aids cooking and browning the meat helps the appearance, by giving a sheen to the surface. Steaks and chops 21/2 cm. to 4 cm. thick should be cooked in moderate heat. A steak 11/2. cm. and 2 cm. thick should be exposed to higher temperature. To test whether it is cooked, the meat should be pressed. If it resists, then it is cooked. The pressure brings the juices to the surface and their colour is a guide.

Hamburgers are usually griddled, i.e. placed on a flat hot surface, with a little fat; and cooked on both sides. The hamburgers are placed on one side and allowed to brown and when the juices rise to the top, they should be turned on the other side and cooked. Only those meats are fried that have a protective coating and binding, e.g. meat croquettes, meat balls, cutlets, scotch eggs, etc.

Judging the Quality

Beef

(1) Fresh meat must be hung to allow it to become tender. The colour darkens after it has been hung.

(2) Lean meat should be bright red, with small flecks of fat interspersed in the muscle (marbled).
(3) The fat should be firm, brittle in texture, creamy white in colour and odourless.

Veal

(1) The flesh should be pale, pink, firm, not soft or flabby.
(2) Cut surfaces should be moist.
(3) Bones in young animals are pinkish, white, porous and with a very small amount of blood in their structure.

Lamb

(1) Carcass should be compact and evenly fleshed, having an even coat of fat.
(2) Lean flesh is firm and of a pleasing, dull red colour and of a fine texture of grain.
(3) Fat should be evenly distributed, hard, brittle, flaky and clear white in colour.
(4) The bones are porous in young animals; as age progresses, they are smooth, white and brittle.

Pork

(1) Lean flesh should be pale pink changing to rose as the animal matures.
(2) The fat is white, firm, smooth and not excessive.
(3) Bones must be small, slender and pinkish.
(4) The fat ought to be white, smooth and not excessive in proportion to the bacon.
(5) The lean meat should be deep pink in colour and firm.

Liver / Offal

(1) The liver of animals should be fresh and not dry.

(2) Should be smooth in texture and not contain tubers in the flesh.

(3) There should be pleasant odour.

Kidneys
(1) Should be fresh.
(2) Certain amount of fat should be attached to the kidneys as it keeps them moist.
(3) The fat should be crisp and not sticky.

Tongue
(1) Lamb or ox-tongue should be fresh.
(2) There must not be an excessive amount of waste at the root end.

Sweetbreads
(1) They should be creamy, white in colour.
(2) They should have a pleasant smell.
(3) They should be fleshy and large in size.

Oxtail
(1) They should be of good size and lean.
(2) There should be no signs of stickiness.

Head
(1) Pig's and calfs head should not be sticky.
(2) They should be well fleshed and odourless.

Lamb Mutton

Lamb is the meat of a young sheep, mutton the flesh of the mature sheep or goat. *Lamb* usually is the flesh of animals not more than 14 months of age.

The average weight of an imported lamb is approx. 32 lbs. (16 kg.), and for mutton, it is upto 50 lbs. (25 kg.). The average weight of an Indian lamb is 10 kg. to 14 kg. For mutton, it is between 20 kg. to 22 kg.

Dissection of Joints

Lambs/Mutton

1. Remove the shoulder.
2. Remove the breasts.
3. Remove the middle neck.

4. Remove the scrag-end.
5. Remove the legs.
6. Divide the best end from loin. (2 loins attached is called saddle.)

The joint that has two legs and the saddle attached, and when it is cooked whole is called a baron of lamb.

Some of the dishes prepared with small cuts of lamb of mutton, are obtained from certain main joints as follows:

Lamb or Mutton Chop

1. The prepared loin is cut into chops approx. 4 cm. thick and trimmed. Grilled or fried.
2. The chops are taken from the leg end of the loin. Trimmed, usually are braised, roasted or fried.
3. It is obtained from the saddle and roasted. The dish, crown roast is very popular. It looks interesting with the paper frills placed on the ends of the bones.
4. The chump end of the loin is removed along with the fillets. The loin is then deboned and thick slices are cut from the meat slantwise. Flattened, trimmed, to cutlet shape and then roasted or grilled.

For Indian dishes, the cuts of meat used are not the same as above. No specific joints are used except for chops. Kashmiri chops, and chop covered with mashed potatoes are popular; for the Kashmiri dish "Kabargah" chops are also used. For kormas, roghanjosh, kaliyas, etc. (Indian curries) any part of mutton is used. Minced meat is used for koftas (meat balls), kababs, shami kababs, seekh kababs and mutton cutlets, etc. When deboned meat is to be used in the form of juliennes or dices, the legs or shoulders are used, e.g. jhalfrezi, mutton do peaza, etc. For biryanis and pulaos, any cut of the meat is used.

Pork: The flesh of a pig is called pork. Most of the pork

meat comes from animals not more than a year old. It generally has more fat than other meats. A suckling pig is about 5 to 6 weeks old.

First quality joints of pig are the leg and loin of pork.

Order of Dissection

1. Remove the head.
2. Remove the trotters.
3. Remove the legs.
4. Remove the shoulder.
5. Remove the spare ribs.
6. Divide the loin from the belly.

Ham: It is taken from the hind leg of the pig, preserved by curing or pickling in brine, then dried and smoked. Ham is prepared from fresh pork meat. Boiling of ham is done by placing it in cold water first and soaking it. Then it is placed in a suitable container covered with cold water and brought to boil slowly, till it is cooked. (Cooking time is 18 to 20 minutes for 455 gms. of ham.) Hams after boiling can be decorated with chaud froid sauce. Gammon is taken from the leg of the pig that has been reared for bacon, and the meat is cured. Gammons are mild and do not keep as long as true hams.

Bacon: Bacon is obtained from the sides and back of a baconer (a pig reared and specially fed to yield bacon). The bacon is acquired by taking the meat and preserving the meat by salting. After the meat has matured, the bacon is sold as "green" bacon. Smoked bacon is very popular and this is done in smoke chambers, which gives it a distinctive flavour, and helps to preserve it for a longer time. The wood that gives good flavour are from the breech tree, or oak tree, etc.

Beef: Beef and veal are got from the bovine animals and they are classified by their ages-

Veal -- Flesh of calf (less than three months of age) which lives on milk.

Calves-- The animals are from 3 to 8 months old.

Beef -- The meat is taken when the animal is above 8 months.

Beef is the most popular of all the edible meats in the Western countries. In India, beef is not very popular and buffaloes are slaughtered in some places and sold as beef.

Beef is the flesh of steers, heifers, cows, bulls and stags. The age and the sex has an influence on the taste and quality of meat.

Order of Dissection

Side: The carcass is divided into two lengthwise. For Hindquarters:

1. The whole side is divided between the wing ribs and the foreribs.
2. Remove the rump suet and kidney.
3. Divide the loin and rump from the leg (topside, silverside, thick flank and shin).
4. Remove the fillet.
5. Divide the rump from the sirloin.
6. Remove the wing ribs.
7. Remove the shin.
8. Bone out the pitch bone.
9. Divide the leg into three remaining joints (silverside, topside and thick flank).

For Forequarters:

1. Remove the shank.
2. Divide in half down the centre.
3. Take off the foreribs.
4. Divide into joins.

Beef Terms

Carcass	-	Whole animal after dressing from the slaughter house.
Side	-	A carcass that has been split down the centre of the spinal column.
Quarter	-	A side that has been sub-divided into two sections as--Forequarters-- Hindquarters.
Corp	-	Forequarters minus the plate, brisket and shin.
Baron	-	A complete back (uncut) which is a pair of sirloins undivided.
Offals	-	Products derived from the process of slaughtering other than the carcass, the edible internal organs, e.g. liver, heart, sweet-bread, head, tongue, tail, brains, kidneys.

Steaks: Steak is a slice of beef varying in thickness, usually taken from fillet, rump or centre fillet. Steaks are tender and have a good flavour and are very popular. They are usually cut into individual portions, that are thick enough to broil well and to develop a crisp well browned exterior, while retaining a juicy interior. Steaks are obtained from different parts of the animals such as:

Entrecote Steaks: These steaks are taken from the deboned sirloin without the fillet. This joint, known as contre fillet, undercut (faux fillet). From this contre, fillet are cut those famous steaks like Minute Steaks (Fr. Entrecote Minute), Sirloin (Fr. Entrecote), Double Sirloin Steaks (Fr. Entrecote Double).

One cm. slices are cut from the flesh of contre fillet. They are thin steaks, flattened and weigh approximately 115 gms. Sirloin steak (Fr. Entrecote). It is a slice of steak, often taken from contre fillet. It is a little thicker than minute steak, 1.25 cm. thick, approximate weight is 150 gms.

Double Sirloin Steaks (Fr. Entrecote Double): It is a steak bigger and heavier than Entrecote steak. It is 21/2 cm. thick and the approximate weight is 250-300 gms.

Porter House Steak and T-bone Steak: Like the Entrecote, it is cut from the sirloin of beef, usually from the best part of the loin, 3 or 4 cm. from where the fillet begins. It is cut or sawn right through including bone, loin and fillet. It is approximately 4 cm. thick and weighs about 650 gms. This is a nerous portion and is large enough for 2 portions. T-bone steak is an individual porter house steak.

Fillet Steaks: Special names have been given to the steaks cut from the different areas of the beef fillet.

Chateaubriand: It is a large steak, cut from the head of the fillet, which has been trimmed of nerves, etc. Approximate weight is 250 gms. to 1 kg. It is 3.5 cm. to 8 cm thick.

Fillet Steaks: It is cut from the upper middle part of the prepared fillet. The steaks are 2 cm. thick and the approximate weight is 100 gms. to 150 gms.

Tournedos: A round barrel shaped steak cut from the lower middle of the beef fillet. It is 2.3 cm. thick and weights approximately 100 gms.

Fillet Mignons: It is prepared from the tail of the fillet, weighs approximately 50 gms. It is small fillet. The ends of tail at one time were used to prepare carpet-bag steaks.

Steaks from other Parts of the Animal

Point Steak: It is approximately 200 gms. in weight and is cut from the triangular part of the rump which is generally 2 cm. thick.

Tartare Steak: Finely chopped fillet of beef, finely

chopped onion, parsley and capers. This is the only steak that is served raw.

The above ingredients are mixed, well moulded into flat cakes, a depression is made in the centre of the cake with an egg yolk placed in it, and served garnished with neat bouquets of onion, parsley and capers all round the steak.

Cooking and Serving of Steaks: Steaks are grilled or sauteed. For thicker steaks, a lower temperature (325°F) should be used for cooking. Steaks could be cooked under done as rare and medium done. The cooking time of steaks varies according to the thickness and size of the steak.

Different garnishes could be served with the steaks. They should be accompanied by a suitable sauce, e.g. sauce bearnaise, sauce diable or chateaubrai and sauce.

A Popular Form: The origin of this popular form of meat food is difficult to trace. It was over two thousand years ago that the Romans enjoyed this delicacy and various forms were known as Tomacina, Circelli, etc. The word "sausage" can trace its ancestry to the Latin salsus (salted) which might suggest that the original type was prepared from cured meats, which could be preserved for a longer time. The ancient town of Salamis provided a place, association for salami and its fame evidently spread, as we now have many types; Italian, French and German Salami, usually with a prefix indicating the locality of origin.

Crepinettes are flat sausages made from fresh pork and wrapped in caul. Sausages can be considered under three main sections, though there will be, of course, some overlapping between the sections.

Composition: Sausages consist of forcemeat prepared with minced pork, poultry, veal and lamb.

Casings: The intestines of sheep, hog and cattle according to the required diameter. The natural casings are available but require a great deal of labour to be processed satisfactorily.

Cellulosic or artificial intestines are made of edible or inedible cellulosic material. They are available in all sizes and are adapted to specific uses. The casings can stand boiling temperature and they are porous and let the smoke go through.

Forcemeat: Pork is used as a basic ingredient for forcemeat stuffing. Pork is minced, ground, chopped or diced into what is called "Forcemeat". Every forcemeat has a binding such as rice, potatoes, panada or fat which holds meat. It is used for the preparation of many speciality items and is also used as stuffings or garnishes for meat, eggs, fish, poultry, game or vegetables.

The composition of forcemeat is made up of 40% dominant flavour meat, i.e. chicken, meat, lambs; 30% pork meat -- it gives flavour and moisture; 30% pork fat imparts tenderness, binds and contributes to flavour.

To this mixture herbs, spices, nuts and seasoning are added. The forcemeat is used to prepare terrines, pates, sausages, galantines.

Fresh Sausages: As they are a perishable product, speed in production and distribution is essential. Pork, beef and veal meat is used. Some of the sausages have cereals added to the meat and seasoning, e.g. chipolatas, bokwurst.

Cooked Sausages: Beef, veal, ham, tongue and pork are used to prepare the sausages. Sometimes a binder is used. The minced meat is cooked and then stuffed into casings. Liver sausages, black pudding, Frankfurters, etc.

Dry Sausages: Pork is used, which is an important

constituent. The Food Regulations necessitated the pork to be heated to a temperature not lower than 140°F or refrigerated not below 48°F as a precaution. Dry curing is done after pickling and then smoked. The dry sausages are salami, mortadella.

Ingredients: Standard quality ingredients should be used, with proper seasoning and binder. Fresh beef and pork have greater binding qualities than the frozen variety. Good quality meat should be used, pork fat is added, if necessary. The internal fats have a higher melting point. The fats should be absorbed by the binder or the excess oil will collect under the skin of the sausage and be detrimental not only to the appearance but also to the taste. A product containing no lactose or milk fat but containing 85% Protein was found and it is known as gervita. By using gervita in sausage, the fat contained therein is in a stable form and the loss on cooking is reduced. Another important aspect is that it swells and minimises shrinkage. The meat used must be finely minced.

Binders: Such as bread, rusk, rice, flour, farina (potato flour) are used in small quantities.

Flavouring and seasoning ingredients used are cayenne pepper powder, paprika, allspice, cinammon, cloves, pistachio nuts, mustard powder, sage, thyme, etc. Salt, vinegar, saltpetre, herbs act as preservatives. The mixing of the ingredients should be thorough to get good results.

Sausage casings are obtained from the intestines of sheep, cattle and pigs, which are carefully cleaned. Cellulose and synthetic casings are being widely used. Different diameter casings are available for the sausages.

The mixture is pumped into the casings. The sizes, different ingredients and processing give different names to the sausages.

Various Types

Frankfurt Sausages: It is a German speciality and is most often used to accompany sauerkraut. They are served hot as hors d'oeuvres with grated radish. They are made from a forcemeat composed of lean pork, trimmed of all fat, minced with bacon. Saltpetre is added to give a slightly pink colour. They are smoked, which gives them the typical yellowish outside colour. German Frankfurters are thus called if they come from Frankfurt of its immediate surroundings.

Strasbourg Sausages: These sausages are made like smoked Frankfurt sausages. They are mainly used for garnishing sauerkraut.

Chipolata Sausages: They are long twisted sausages made in very small sized casings. These are divided into small sausages and can be grilled, sauteed in butter or poached in white wine. They are made of pork meat and rusk is the binder with pepper, salt, pimento, nutmeg, thyme and cayenne pepper. When cut, it is slightly pinkish colour and is served with roast turkey and Rognons saute turbigo dishes.

Mortadella: It is a larger oval shaped type of sausage of Italian origin. The Mortadella of Bologna has the highest reputation. Mortadella is also made in France. This is a smooth sausage prepared with pork and beef with pork fat, garlic, pistachios, cardamom, cloves, salt, pepper, etc. It is filled in beef bladders and smoked.

Salamis: They are smoked and usually served sliced. They are medium sized sausages of beef and pork with garlic and salt, peppercorns, etc. The texture varies from coarse to very finely cut meat.

The most renowned is the Bologna. The different salamis are:

Belgian salami: It consists of beef and pork in a distinctive white rasing- without garlic.

Italian salami: It consists of pork and beef with rice like speckles f fat It is well seasoned with wine and garlic.

Danish salami: It is made up of pork and veal, bright pink in appearance with a distinctive salty flavour and slight garlic flavouring.

Black Pudding: One of the earliest cooked sausages consisting of blood, oatmeal, onions, pepper, rice flour, pimentos, salt, mustard powder, etc. They are black in colour.

9

Poultry Dishes

Poultry is the name given to domestic birds, specially bred to be eaten. It includes the different varieties and sizes of chicken, turkey, ducks, geese and guinea fowls. Drawn poultry loses approximately 25% of its original weight. The word 'Volaille' in French means poultry, but when it appears in the menu, it means chicken.

When the size of the bird has been chosen, the next point is to recognise signs of young and fresh birds, as it helps in the preparation of chicken dishes.

Signs of a Fresh and Young Chicken

(1) Soft fluffy feathers should be all over the body, especially concentrated under the wings.

(2) The quills of the feathers should be easy to pull out.

(3) The skin should be clear and smooth with no bluish tinge spots.

(4) The feet should be supple with smooth overlapping scales, and if fresh, moist.

(5) Combs and wattle (the skin hanging near the beak) should be small and well developed. The comb being bright red.

(6) Conformation or the shape of the bird is important, and the bird should be free from deformity.

(7) The eyes should be prominent and clear.

(8) It should not have a marked unpleasant smell.

Conditions in Different Birds

Chicken and Fowl: The breast should be plump, firm flesh, pliable breastbone.

Turkey: The breast should be plump, the wattle red. Under four and half kg should not be chosen or the proportion of bone to flesh will be great.

Duck: The breast should be plump. The bill and feet are yellow in young birds, but they darken with age. The webbing of the feet should be smooth, tender and easily broken and the under bill soft and pliable.

Goose: The legs should be smooth and the breast plump. Feet should be yellow and smooth, and webbing should be easily torn.

After slaughtering the birds, they should be held at a cool temperature, but, not frozen, for 4 to 8 hours. Turkey should be held for 12 hours in cold conditions.

Food. Value: The flesh of poultry is more easily digested than that of butcher's meat. It contains proteins and is therefore useful for building and repairing body tissues and providing heat and energy.

Cooking Methods: The methods used for poultry depend upon the size of the bird and the nature of the flesh. Some birds are young, tender and lacking in fat, others are tough and require moist heat for tenderizing. Older fowls have richer flavour to impart to soup and creamed dishes than young chickens. The characteristics of the flesh have led to special methods of preparations for the specific classes of poultry. Young birds are cooked by grill, tandoor or roasted as they are tender. Chicken pies or chicken dumplings, and pakodas are made with fat fowls. Matured birds are cooked by moist cooking, spices are added when flavour is lacking and are enriched with fat such as fricassee, chasseur, kormas, curries, etc. Lean young birds are fried as it enhances Palatability.

Cleaning

(1) Remove any pin feathers with a small knife.
(2) Singe (place it over an open flame to burn the hairs) but do not scorch.
(3) Cut the head off and slit the neck skin lengthwise on the back.
(4) Cut off the neck as close to the body.
(5) Put the forefinger from top and loose the crop, intestines.
(6) Make a slit near the vent (rectum) and loosen the internal organs with the fingers.
(7) Draw out the innards being careful not to break the gallbladder.
(8) Clean the gizzard, keep the liver, neck and heart.

Cutting for Sauteing: This is the Jointing of a chicken at the raw state. The cuts are made mostly through the natural joints to produce two wings, two pieces of breast meat, two winglets, two drumsticks, and two thin pieces. The carcass can be used for stocks.

When cut this way, a chicken will lend itself to many preparations, i.e. fricassees, blanquette, saute, curries, pies, tandoor chicken pieces.

	English Cut	*French Name*
1.	Drumsticks	Pilon de Cuisse
2.	Thighs	Gras de Cuisse
3.	Winglets	Aileron
4.	Wings	Aile
5.	Breast	Poitrine or Blanc
6.	Carcass	Carcasse

Trussing

For Roast: The legs are blanched in boiling water for a few seconds and the scales removed with the aid of a cloth. The claws are trimmed. To facilitate carving, the wishbone is removed. The bird is then trussed, using two strings. The trussing needle is a strong steel needle about 6" long with a large eye. Thread the trussing needle and pass it through both the thighs and then through the wing, the neck, which has the Skin covering the neck opening, and then through the other wing and tying both the ends of the string securely. The winglets should be crossed over the breast. A second string is passed over the legs and under the white breast meat and secured leaving the prepared legs protruding beyond the rear of the chicken.

Stuffing: Stuffing is put inside from the neck and between the skin of the breast and the breast of the chicken. This makes the chicken look plump. Stuffing for murgh mussallam is done in the hollow from where the internal organs have been removed and cleaned.

Preparation of Tandoori Chicken: The neck is cut off and the skin is removed, internal organs are removed. Slits are made in the flesh on the thighs and breast so that heat could penetrate easily. The chickens are marinaded in a mixture of beaten curd (yoghurt), lime juice, orange colour, garlic and ginger paste, Kasuri methi, salt, pepper and garam masala (mixed spices) powder.

For Grilling: Remove the wishbone and the legs from the first joint. Cut through the backbone, open out, remove back and rib bones, flatten slightly.

Supreme: This is the wing and half of the breast of a chicken, or the whole of the white meat from one side, with the trimmed wing bone attached. The white meat of one chicken yields two supremes.

Ballotine: The boned-out raw leg of chicken which can

be stuffed, braised, stewed or sauteed according to the recipe.

Accompaniments and Garnishes: For roast chickens, game chips, watercress, roast gravy and bread sauces are served separately.

Grilled chicken-- Fried straw potatoes, piquant or devilled sauce. *Tandoori chicken--*Onion rings, lemon and green chillies.

Chicken korma-- Chopped green dhaniya and fried cashewnut slices and silver leaf. *Chicken pakodas--* Garam masala powder sprinkled on top, mint or sonth ki chutney. *Chicken curries--*Chopped green dhaniya and juliennes of ginger *Dishes of Chicken*

Chicken Maryland: 4 pieces are taken from the chicken and leg is deboned (2 breasts,, and 2 legs) coated with seasoned flour, dipped in eggs and bread crumbed and fried. Served with grilled bacon rashers, corn galettes, banana fritters, and fried parsley.

Chicken Ballotine Chasseur: Deboned chicken, stuffed and tied to maintain shape. Saute the chicken, add mushrooms, white wine and brown sauce tomatoes, and cook on top of the stove. Then add chopped tomatoes and sprinkle chopped parsley and fried heart-shaped bread croutons.

Chicken Mousseline: The word mousse means 'a finely pounded substance to which cream has been added'. This can be served hot or cold. It consists of finely minced chicken, cream, egg whites, seasoning and nutmeg moulded and poached. Served cold.

Chicken Salad (Salade de Volaille)	Roasted chicken, hard-boiled eggs, tomatoes, gherkins, pineapple, onions, lettuce, olives, vinaigrette or mayonnaise, lettuce and parsley. (Mix all ingredients except olives and parsley and lettuce with dressing.)

Chicken Galantine
Chicken stuffing--(Veal, pork, ham, pork fat, larding bacon, truffles, pistachios, spices, salt, brandy and eggs.). Remove the bones from the back of the chicken, by making a slit and working round the bony structure. On cloth, place the deboned chicken, place a layer of stuffing, placing strips of fat, bacon and nuts in between. Roll like a sausage and tie the ends securely and poach in chicken stock for $1^1/_4$ hrs. Remove, tighten the ends and again cook for 15 minutes. Drain, cool and unwrap the cloth, decorate with clear chicken aspic.

Chicken Chaud Froid
Spring chicken, supreme sauce, chicken aspic, truffles, court bouillon. Use for decoration: tomatoes, lettuce leaves, curry leaves, etc. Truss the chicken and poach slowly in court bouillon (15 minutes per 450-500 gms.) Cool in cooking liquor when done, remove skin, coat the chicken in supreme sauce + aspic and cool it. Decorate with truffles, tomato and curry leaves.

Chicken Saute Viennoise
(Poulet saute viennoise Fr.)
Egg and crumb the deboned chicken pieces, cook in butter and serve with lemon slices, chopped parsley on top, mask with brown butter and serve.

Chicken Florentine
(Supreme de Volaille Florentine Fr.)
Poach dress on a layer of cooked spinach leaves coat with mornay sauce and glaze.

Chicken Stanley
(Poularde Poche Stanley Fr.)
Stuff chicken with rice, truffles, mushrooms, onions and butter. Truss and poach in chicken stock, remove chicken, reduce liquor, add cream and place chicken on rice bed (rice and mushrooms) and coat the chicken with sauce and serve.

Chicken Laperouse
(Poulet saute Laperouse Fr.)
Saute chicken in the sauce pan, add brandy, cream, mushrooms puree and yolk of eggs, garnish with artichoke bottoms containing mushrooms puree and gratinated.

The other dishes that are made out of chicken are: Quenelles, medallions, cutlets, chops, croquettes, salpicons, salads, bouchees, etc.

Various Cold Dishes

Pates - - Forcemeat is put in various types of utensils and is not covered while cooking. Actually the word pate is a dish prepared of meat or fish enclosed in a pastry or a dough and baked in the oven. A pate is a savoury pie and can be served hot or cold. Liver pates are made of liver. Fat seasoning finely minced and cooked and passed through strainer to get a smooth paste and moulded.

Terrines- -The terrine is made up of forcemeat of any type of fleshy items such as pork, veal, beef or-lamb, poultry, game and seafood. The coarse forcemeat is moulded in earthenware utensils from where it gets its name. Covered with a light fitting lid during the cooking process and served in the same mould. Terrines are always cold.

The Art of Carving: The art of carving in front of the guest is difficult to teach by theory or with the help of diagrams. However, an attempt has been made with the aid of diagrams.

Practice, good tools, careful preparation of the joint before cooking, and a knowledge of the bone structure is essential.

Selection of Tools: For most joints, a knife with a blade 10" to 12" long and about 1" wide is required.

For poultry or game, a knife with a blade 8" long is more suitable For ham, a long thin flexible knife is preferred.

A carving fork is needed to hold the meat and serve the slices.

A steel is necessary to set the edge of the knife. The manner of holding the knife to set it on the steel is important; it should be laid at a low angle (nearly flat), and it should be drawn downwards in a semicircular motion on both sides of the steel.

Preparation of Joints: The preparation of joints before cooking is most important in carving at the table, if the carving is to appear effortless as it should, and not a tug of war.

As bones cannot be cut through at the table, any bone which makes carving difficult should be removed before cooking, or should be sawn through so that it can be removed easily after cooking.

In this respect, the collaboration of the butcher is required. It will of course be necessary to inform the butcher of your requirements.

Hotels and restaurants, having a larder chef, prepare their joints to suit the requirements of their particular establishment, thereby ensuring the maximum possible economy.

Bone Structure: Knowing where to expect bones is most important in carving. This refers to those bones that are part of the internal structure of the joint, and that could not be removed without changing the shape of this joint, i.e. the difference between a shoulder of lamb with the bone in a boned and rolled shoulder.

A joint is useful to locate the bones, but it will be necessary to study it beforehand.

The following methods of carving do not represent the

only possible way. They are those that are generally accepted and practised.

Chicken: With wishbone removed before cooking:

1. Remove leg by cutting around skin A, stick a fork at XXXX and pull outwards, severing joint. Cut leg in two at B.
2. Carve the wing off at C, cutting through the joint.
3. Repeat for other leg and wing.
4. Cut on either side of the breast- bone, in order to remove the meat D. in two long pieces.

   Without wishbone being removed: 1,2,3 as above.
5. Lay bird on its side, cut through X-X, and pull whole breast off: Divide in two at D, or cut at D first, lift off the meat from the bone.

Duck

1. Remove leg as for chicken.
2. Carve the wing C off, as for chicken but shorter.
3. Cut the breast in thin slices, holding the knife slightly inclined outwards.

Ducklings, having little flesh on the breast: It will be easier, after removing leg to cut away wishbone (if not previously removed), make incision along breastbone and lift wing and breast in one piece, slicing in two or three afterwards.

A small turkey can be carved without removing the legs.

1. Cut leg first at A, then under drumstick at B, slicing inwards, next carve the flesh off the thigh at C.
2. Carve the breast diagonally, starting at D, proceeding towards the wing E.

3. Serve a slice of white meat and a slice of black meat.
4. When all the meat is carved off the left side of the turkey, turn it round with the leg towards the carver and proceed in the same manner.

A large turkey can be carved without removing the legs, but it will be found easier to remove and carve them separately.

Partridge: Cut along each side of the breastbone, push knife sideways and downwards to detach the meat from the carcass, serve through shoulder and thigh bone, and serve leg, or breast and leg wing in the piece.

Guinea Fowl, Grouse, Pheasant, etc. similar to chicken.

Pigeon, Plover, Woodcock, Snipe, Quail, etc. are small birds normally served whole.

Hold the knuckle bone with a serviette (or paper frill).

Make an incision above the knuckle at A.

With the knife slanting towards the knuckle, cut two or three slices, at B according to size of joint.

Then proceed to cut alternate slices from C. Finish the carving by slicing the fillet D.

10

Vegetable Dishes

Vegetables play a very important part, as more and more people are becoming vegetarians and vegetable's are widely used in cookery. They are served in various forms, as accompaniments, garnishes, salads, hors d'oeuvres, bhajees, sookhi; sabzi, curries, raitas, etc. There is a lot of repetitive work entailed in the preparation and cooking of vegetables. Vegetarian cooking demands a good knowledge of the principles of cookery and a high degree of skill. During the preparation of all kinds of fresh) vegetables, there can be excessive amounts of waste unless it is efficiently and economically carried out. All vegetables, particularly the green ones, are best when taken straight from the garden.

Vegetables could be classified in two ways for culinary purposes:

Parts of a plant:

(1) *Roots*: Carrots, horseradish, beetroot and turnips.
(2) *Bulbous roots:* Onions, shallots, leeks, garlic.
(3) *Tubers:* Potatoes, Jerusalem artichokes, kachaloo.
(4) *Flowers:* Cauliflower, broccoli, globe artichokes.
(5) *Leaves:* Cabbage, lettuce, spinach, watercress, sarson.
(6) Stem: Celery, rhubarb, seakale, chicory, asparagus.

The other way is to broadly classify such as--Root vegetables. Leafy vegetables. Other vegetables.

(7) *Fruit:* Cucumber, tomatoes, beans, peas, pulses, pumpkin, tin das, lauki.

Nutritive Value

Except for the dried vegetables such as dried legumes, vegetables have a high water content, which ranges from approximately 70 to 95%. Of the other foodstuffs present in vegetables, carbohydrates rank first in quantity. These may be present in the form of starch, sugar, cellulose, pectic substances. The quantity depends upon the part of plant from which the vegetable comes and the degrees of maturity--ripe or unripe. Starch is the chief nutrient of roots and tubers; potatoes have the highest percentage. Sugar is highest in beetroot and it is also rich in carrots and turnips.

Cellulose, the fibrous framework or roughage becomes coarser and tougher as the vegetables mature. In some cases it is too tough to be eaten raw compared, e.g. to lettuce and cabbage. If the dishes are well chosen and eaten raw or cooked, they contribute vitamins and minerals to the diet. They are provided by vegetables, as they contain relatively large amount of vitamins and minerals as compared to other foods. Besides nutrient components, vegetables have certain other materials (components) which have no food value, but are of great concern in vegetable cookery. They have an effect upon flavour, colour, and general palatability of the cooked product. These materials are-- flavouring substances, plant pigments and tannings.

Flavouring Substances: Many volatile and non-volatile organic acids contribute to flavour. But volatile aromatic oils are chiefly responsible for characteristic flavour and odour. Some are present in a raw state and others develop when food is cooked.

Pigments: The attractive and varied colours of vegetables are due to the presence of pigments distributed in plant tissues like chlorophyll (green) and carotenoid pigments (red and yellow).

Tannings: Complex organic compounds are widely distributed in plants. They have astringent properties. Tannings of tannic salt form greenish purple compound, which upon oxidation, turns brown.

Selection of Vegetables: The quality of root vegetables is determined by their firmness and absence of blemishes. The skin should be smooth and not wrinkled for carrots. The potatoes should be of regular size and free from earth. The vegetables should be heavy for their size. Onions should have thin shiny skin and no signs of sprouting. Green vegetables should be crisp, have a fresh, green colour appearance. If the ends of the stems are broken, they should snap off easily and the leaves should not be limp. Cabbage should be compact and be heavy for its size. Cauliflower should be white, the flowerettes well formed and the leaves crisp. Spinach should be crisp with slender stalk ends, and the leaves large in size. Tomatoes should be bright red, firm, regular shaped with shiny skin. Beans should not be stringy and should break crisply under pressure. Pea pods should be plump. Brinjals should be light for their size, having a shiny and smooth skin. Lady's finger should snap if the ends are broken. Carrots should have a shiny skin and regular shape.

Vegetable Preparation: The nutritive value which vegetables contribute in the diet should be conserved as completely as possible. Most of the vegetables grow near the ground and get contaminated by dirt and sprays, etc. They must be properly washed especially those which are to be are peeled thickly to remove the inner skin, which is tough and tasteless. Green vegetables should be washed well before cutting so that the grit is thoroughly removed. Potatoes and carrots are washed first and thinly peeled. Vegetables are cut into different shapes such as dices, batons, juliennes (matchstick) roundels, triangular shapes, and they may be chopped, minced, sliced and grated, etc.

Preserving nutrients during the preparation and cooking of vegetables is important, and following are the ways to preserve nutrients:

(1) Do not let vegetables soak in water unnecessarily.
(2) Cook or bake vegetables in their skins.
(3) Keep air-oxygen away from vitamin rich foods by covering foods while cooking; not stirring food unnecessarily as it mixes air; cutting vegetables just before cooking, and serving food as quickly as possible when it is cooked.
(4) Use as little water as possible in cooking, so as to preserve vitamins, minerals, flavour, colour and texture of food.
(5) Baking powder should not be added while cooking vegetables.
(6) Do not expose food, milk in particular, to light.
(7) When preparing frozen vegetables, place directly into boiling water.

Effect of Cooking on Vegetables: Vegetables are cooked for various reasons such as improving their palatability and digestibility. In the case of certain food, flavours are improved and preserved longer by cooking.

Certain changes occur during cooking as:

Cellulose: It disintegrates and becomes soft and mastication is facilitated.

Flavour: Flavour is affected during cooking, and over-cooking adversely affects flavour. A covered pan increases the intensity of flavour.

Pigments: The colour pigments of fruit and vegetables are chlorophyll, carotenoids, flavones and anthocyanins.

Green vegetables because of their chlorophyll when cooked, for a long period at high temperature, tend to lose their colour. The green coloured vegetables are cooked in an alkali medium, to preserve the loss of colour, but certain nutrients are lost--carotene of carrots while cooking is not lost: in fact, it brightens the colour. Anthocyanins, in red and purple coloured vegetables like red cabbages, beetroots which are highly soluble in water and high temperature, bleaches them. Flavones are found in white onions, cauliflower, etc. Their white colour could be improved by cooking them in water to which lime juice or vinegar has been added.

The changes in carbohydrates take place as the heat partially or completely gelatinises starch. Swelling of the starch grains occur during cooking of vegetables and, in some circumstances, the sugar caramelizes.

Proteins are negligible in vegetables except for legumes: heat coagulates completely or partially.

Cooking losses occur through the solvent act of water or the chemical decomposition which may be caused by the reaction of the cooking water or by heat. Other common cooking losses are through oxidation, or by volatilization.

Vitamin C is lost because of high temperature. Vitamin A, Vitamin B2 and Niacin are heat stable.

Cooking Methods: Common methods used for vegetable cookery are boiling, frying and stewing, though methods are also used such as steaming, braising, grilling and baking.

Boiling: Green vegetables such as spinach, cabbage, peas and beans, etc. are placed in boiling salted water and rapidly boiled because it helps to conserve the colour and flavour. Root vegetables are first placed in cold salted water and brought to boil and simmered till tender. Care must be taken to ensure that they are not over-cooked. The cooking liquor should be used and not thrown away. Blanching green vegetables, retains a bright colour, prevents discolouration, retains texture, stops further cooking and retains nutrients.

Braising: It is mainly for celery, leeks, cabbage, onions, and lettuce. They are blanched and placed on a bed of cut aromatic vegetables and moistened with stock and braised in the oven, e.g. celery, spinach, etc.

Steaming: Usually potatoes are steamed. Other vegetables also can be steamed in a pressure cooker.

Frying: Vegetables cut into various shapes, can be fried and they can be dipped in a batter and fried.

Grilling: Tomatoes and mushrooms are grilled.

Baking: Usually the vegetables are baked in their skins, often on a bed of salt. Brinjals for bharta are baked. Sweet potatoes and potatoes, also can be baked.

Continental Vegetables

Artichoke: It is cultivated for its flowering head--used for highclass hors d'oeuvres, garnishes and a vegetable course. The fleshy base is the part usually eaten. Tips are also eaten.

Asparagus: Agenteuil asparagus is very popular. The part that is eaten is the young shoot. It is usually served with butter or sauce and is available in India. It is used as a vegetable course or as a garnish. It is delightful when cold and served with vinaigrette of cut and reheated in butter as a garnish.

Endive: Belgian chicory (endive) is the best. Green endive leaves are very bitter. Therefore they are blanched to pale yellow colour to make them palatable. Cold preparations are served as a hers d'oeuvre. It is also used as a vegetable and for salads.

Leeks: Leek resembles *hara pyaz* (spring onions), but are bigger and have a lighter shade of green. They are used as *a* hors d'oeuvre. It is also used as a vegetable and for salads.

Celery: A vegetable that has a good flavour and is rich in mineral salts and Vitamin C. It is used as a hors d'oeuvre, soup, in sauces, garnishes, salads and served with cheese course.

Kale: The leaves are purplish or light red in colour. It resembles a wild cabbage. It is boiled and has a nutty, slightly bitter flavour.

Parsnip: Widely used in England and it is like turnips and yellowish in colour. It is used in soups, as garnish or as a vegetable dish.

Swede: Often called the Swedish Turnip. It has yellowish coloured flesh. Mainly used as a vegetable.

Cardoon: Plant common to the celery of which the stalk leaves are eaten as a vegetable. Stalks are crisp and tender. It is similar to celery from which it takes its culinary uses.

Indian Vegetables

Aubergine (Eggplant, Baigan or Brinjal): It is available all round the year. The colour is from light purple to blackish purple. Used as a vegetable course and as a garnish. Used widely for Indian dishes, in the form of curries, barva baigan, kachadi {slices dusted with atta and shallow fried), in sambar, bhartas, etc.

Cabbage (Bandh Gobi): They are rich in Vitamin C, widely used in Western menu for salads, as a vegetable and in soups. The Indian dishes are bhajees, foogaths, thorans and for pulaos, etc.

Cauliflower (Phool Gobi): It is a high-class vegetable according to Western standards and is used in soups, as a garnish and as an hors d'oeuvre. The Indian dishes for cauliflowers are pakodas, tahari, cauliflower sabzi, bhujia etc.

Cucumber (Kheera): It should be tender without seeds. Used for hors d'oeuvres, cold soups, indispensable with salmon and fish dishes as garnishes. Doria is a keyword here, also used for cooked vegetables and savouries. In India dishes, cucumber is used for kachumbers, raitas, etc.

Lettuce (Salad ka Patta): There are different types of lettuce--Romaine or Cos lettuce. Widely used as hors d'oeuvres, soups, garnishes, hot vegetable, accompaniment and salads.

Marrow (Lauki): Member of the gourd family. It has a high water content and has little food value. Mainly used as vegetable, raitas, halwas, etc.

Peas (Matar): It is a popular vegetable used in hors d'oeuvres, soups, garnishes, and as a vegetable dish. It is also used in salads. The Indian dishes are matar chawal, matar keema, matar paneer, sookha matar, etc.

Pumpkin (Kaddu): If the pumpkin is green, then it is prepared with the skin and if it is yellow, the skin is removed. It contains 90% water. It is used as a vegetable, pumpkin pie or halwa.

Spinach (Palak): It is a nutritious green leaf vegetable. Extensively used as a garnish and as a vegetable, mutton palak, pakodas, etc.

Tomatoes: Very popular vegetable and used in various

forms--purees, sauces, or paste. It is extensively used in hors d'oeuvres, soups, fish dishes, sauces, vegetables, salads, savouries, curries, chutneys, pulaos, raitas, etc.

Capsicum (Shimla Mirch): It has very good flavour and is used in hors d'oeuvres, garnishes, stuffed or sauteed as a vegetable, and salad. In the West, there is a variety of the capsicum family. The large ones are called pimentos, which could be green or red. When dried and powdered, it is called paprika powder. The smaller ones, the green and red peppers are more pungent and are called chillies. Used for sauces, stews, and pickles, curries, subzi, etc. When dried and ground, it is called cayenne pepper.

Rhubarb: It is a vegetable and consists of a leaf stalk with a pleasing flavour. Used mainly for stewed pies and puddings. In India, rhubarbs are Pickled.

Mustard Leaves (Sarson ka Sag): Extensively used in the north of India. For the Punjabis, it is a speciality dish.

Horse-Radish (Mooli): It has a pungent flavour, is used in hors d'oeuvres, decoration for cold buffets, salads. It is used in India for bhajees, parathas, etc.

Turnips (Shaljum): It is a common root vegetable. Main use of the turnip is as a flavouring vegetable and for garnishing and as a subzi.

Beetroot (Chukander): The sugar-beet goes to sugar manufacturers, the common beetroot to the kitchen. Leaves can be used to prepare vegetables and salad.

Carrots (Gajar): Contains sugar, mineral salts and certain vitamins. It is used for hors d'oeuvres, soups, sauces, garnishes, hot vegetables and salads, gajar ka halwa, pulaos, etc.

Onions (Pyaz): These are available all the year round. Probably the most used vegetable in the kitchen.

Potatoes (pommes de terre, fr.): Potatoes play an important part in the general diet of the population today. No menu seems complete until potatoes have been used, in the form of hors d'oeuvres or as an accompaniment to the main dish. For Indian menus, potatoes are used in the form of curries, subzis, raitas and taharis, etc.

Chemically potatoes are largely composed of water and starch, it is estimated that 99 gms. of potatoes yield approximately 80 calories. The sugar content of potatoes ranges from slight traces to as much as 10% of the dry weight of the potato. The two chief factors influencing sugar content of potatoes during storage are variety and temperature.

There are two main seasons for potatoes, the so called "Earlies" or "Pahadi" and the "Main crop" or "Deshi" potatoes. Therefore the potatoes are available all the year round.

Pre-preparation of Potatoes: Those potatoes should be selected that do not have deep eyes, irregular shape and do not have any discolouration during peeling and trimming. The green colour frequently seen immediately beneath the skin indicates the presence of chlorophyll. Much greening in potatoes in undesirable because of its association with a bitter taste. The potatoes with thin skin should be scrubbed, scraped and cooked in their skins, and the ones with thicker skin should be peeled. To prevent the potatoes from discolouring after they have been peeled, they should be kept in water.

Raw potatoes are shaped or cut into different ways: big or small dices, potato balls, matchstick strips, quarters, roundels, drum shaped rectangular strips, etc. Some of the shapes and sizes are:

Methods of Cooking: Potatoes could be fried, baked, roasted, steamed, boiled and sauteed.

The method of cooking could be selected according to the potatoes, as there are two different kinds of potatoes. The mealy or floury potatoes when cooked burst immediately on application of a fork and break to pieces and crumble. Baking, steaming in jackets, boiling and mashing are the methods of cooking used for this type of potato. The other variety is close and waxy and the flesh is firm, which could be cut into different shapes, as it will retain its shape when cooked. Frying and sauteeing are the common methods used for this variety.

Potato Dishes

Baked Potatoes: (Pommes au Four Fr.) Wash and wipe equal size potatoes, place on a bed of salt and bake in the oven. When cooked, make slits--cross and open by given pressure on either side. Place a dot of butter and sprig of parsley and serve hot.

Arlie Potatoes (pommes Arlie): Bake the potatoes as above, and when baked, cut the potatoes into two, lengthwise and scoop out and mix with chopped onions, butter and cream and pipe the mixture into the jackets, sprinkle with grated cheese and gratinate.

Macaire Potatoes: Bake the potatoes and mash them, season, add butter, make the balls and flatten, dust with flour and saute in butter.

2nd Method: Put the mashed potatoes in Pommes Anna mould and cook with clarified butter.

Byron: Prepare as above, hollow the top, sprinkle with Chester cheese, cover with cream.

Robert: To Macaire potato mixture, add chopped onions and chopped hard-boiled eggs, and cook the same way.

Bhurta: Bake/roast potatoes on gas or charcoal. When cooked, remove the skin and mash the potatoes. Heat butter, and chopped onions, green chillies, tomatoes, saute and add seasoned potato mixture and saute.

Bhurta Raita: Mix the above mixture into beaten curd (yoghurt), correct seasoning and sprinkle zeera powder (cumin) on top. Steamed/Boiled

Steamed Potatoes: (Pommes all anglaise, Fr.) Shape the potatoes, cook in a steamer and season.

Menthe: Same as above, add leaves of mint.

Maitre d'hotel: Same as cream potatoes with the addition of chopped parsley.

Jacquettes Potatoes (Pommes Jacquette): Wash and cut the skin in a spiral method and cook in steamer.

Parsley (Pommes Persillies Fr.): Same as anglaise, season, saute in melted butter and chopped parsley.

Vapeur: Same as steamed potatoes.

Cream (Pommes Creime): Cook the potatoes in water, peel and slice, moisten with milk, add butter, season and bring to the boil, add cream before serving.

Shallow Fried/Saute

Allemande: Cook potatoes in their jackets, peel and cut into thick rounds and well buttered.

Brune: Cook potatoes, peel, chop, shallow fry in butter till crisp.

Provencals: As sautee with addition of chopped garlic to the butter at last minute.

Lyonnaise: Slice potatoes and saute them, and then saute the sliced onions and mix together, season, place in a tray and finish in the oven.

Saute Acru: Saute 1/8" dices of potatoes, finish cooking in oven, sprinkle chopped parsley, serve hot.

Aloo ki Chat: Parboil potatoes and shallow fry, sprinkle chat masala and lime juice and toss.

Tikkies: Parboil potatoes, mash and grind, add seasoning, prepare small flat cakes and shallow fry till golden brown on both sides and serve with tamarind chutney.

Oven Cooked

Anna: Potatoes should be cut into cylindrical shape, sliced in rounds, wash and dry and place the slices in layers in an Anna mould (round pan with lid sliding in the slot with handles on either side) with clarified butter, till the top of the pan, compress and close the lid, cook in the oven. When half done, turn for colour on both sides.

Boulangere: Slice potatoes and onions and saute, season, add stock, cook in the oven.

Chateaux: Turn the potatoes to the shape of a large olive, blanch, saute in butter and place in the oven for finishing the cooking.

Cocotte: Same as chateau, but smaller in size.

Fondates: Turn the potatoes into drum shape (egg size), saute in butter and place in a tray, add stock and seasoning and finish the cooking in the oven.

Flamende: Cook button onions turned carrots cooked, mix with Potatoes.

Delmonica: Cut dices $^1/_{2}$, place in a tray, add milk, sprinkle bread-crumbs, dot with butter and place in the oven for cooking.

Hongroise: Chop onions, saute in butter and sprinkle paprika, add tomato concuss and large roundels of potatoes, add stock, cook and then place in oven and finish off cooking. Sprinkle chopped parsley and serve hot.

Deep Fried/Shallow Fried

Straw Potatoes: (Alumette Fr.) Cut potatoes, matchstick shape, fry in deep fat, add salt just before serving.

Bataille: Cut in dices, deep fry.

Bordelaise: Deep fry 1/2 dices, add fried finely chopped garlic, add chopped parsley before serving.

Frites: cut potatoes 2" length and 1/2 square. Deep fry. *Chips:* Cut slices of potatoes, very thin, deep fry till golden brown. *Mignonette:* Cut potatoes double of matchsticks and deep fry. *Faille:* Cut potatoes in large juliennes and deep fry. *Parmentier:* Cut 1/2 dices and deep fry.

Pont Neuf: Cut potatoes 2 1/2 length, 1" width (regular size) and deep fry (shaped like a ruler).

Souffles: Trim potatoes in a rectangle, cut one-eighth of an inch thick, blanch in a moderately hot fat and drain, then put it in hot fat to effect puffing. Dry and strain.

Puree/Mashed Potatoes

Berny: Mash potatoes and egg yolk, butter, nutmeg and seasoning, stir well till it leaves sides of the pan (appareil duchesse), add chopped truffles, shape like an apricot, egg and coat with sliced almonds and deep fry.

Brioche: Prepare appareil mixture as above, prepare flat rounds and place a small ball of mixture in the centre, egg wash and gratinate.

Croquettes: Prepare appareil duchesse (mashed potatoes, nutmeg, butter, egg yolk, seasoning). Make cylindrical shape (2") egg and bread crumb and deep fry.

Duchesse: Pipe the apparel mixture into rosettes, brush egg and gratinate.

Marquise: (1) Take Duchesse mixture and mix with thick tomato sauce and shape like flat cakes.

(2) Take Duchesse mixture and pipe two three rounds leaving a hollow in the centre, put sauteed tomato concasse in butter in the centre, brush with egg and gratinate.

Gratinees: Mash potatoes, add nutmeg, milk, butter and seasoning, heap in gratin dish, sprinkle cheese and dot with butter and gratinate.

Mousseline: Mash potatoes and add whipped cream and mix well. Serve hot.

Indian potato dishes are: Aloo dum, aloo dum Banares, aloo dum Bhojpuri, aloo ki chat, kofta, bhajee, raita, aloo ki tahari, aloo ki roti, etc.

Oven Cooked

Anna: Cut potatoes in cylindrical shape into slices 1/16" thick. Put butter in an Anna mould.

Arrange slices overlapping the bottom lid slides in sides. Season and fill the mould up to the lid. It should be compressed and put in the oven and cooked. When cooked, remove the lid, clear the top by putting it back in the oven. Demould and serve.

Voisin: Same as Anna after the mould has been lined, add grated Parmesan cheese. Cook and serve as for Pommes Anna.

Darphin: Proceed as for Pommes Anna using coarse julienne of potato in place of sliced potatoes.

Cheese: Cheese is one of the by-products of milk. It is prepared from either cream of milk, or cream and milk mixed together. This is soured and coagulated with rennet or acetic acid. The whey is run off and the curd is fermented and cooked. Different degrees of heat and pressure produce different types of cheeses. The harder and longer the cheese is pressed, the harder will be the texture. The degree of pressure denotes the type of cheese required. Cream cheeses are subjected to light pressure. They are then left to mature and the timing of this produces mild or strong cheeses, as they ripen with age. Cheeses range in consistency from soft creamy substance to an almost rocklike hardness. Cheese is made from cow's, goat's, or ass' milk. Some cheeses are flavoured with herbs or spices.

There are four main types of cheeses used in the kitchen, Parmesan (Italy), Gruyere (Switzerland), Cheddar (England) and Amul (India). These have a strong flavour which is not lost in cooking.

When cooking cheese, it should not be over-cooked, as it is liable to become very indigestible and is liable to "split" (the fat separating from other ingredients). It should always be combined with starchy foods as this aids the digestion. Cheese also becomes rubbery after over-cooking. Low temperature and short cooking is recommended.

Cheese is a very valuable food. It is roughly j protein, $^{1}/_{3}$ fat and $^{1}/_{3}$ water. It has a large calcium content, and also contains vitamins A and D. Cheese is available in natural or processed forms.

Processed cheeses are made by grinding the cheese and then cooking in a vacuum cooker and pouring into moulds and then wrapping in tin foil.

Blue cheeses are made by pressing the cheese, then skewering with copper wires and introducing a type of Penicillin mould into the cheese.

When storing cheese, it should be kept in a cool dark

place and should be wrapped in grease-proof paper to prevent drying. If it is kept at a temperature which is too cold, it will dehydrate and if put in a warm place, it will "sweat". It should be stored away from foods or it will absorb flavour.

Cheese is served in salads, sandwiches, canapes, hors d'oeuvres. Sometimes they are served at the end of a meal. They are also served as a garnish for soups. Paneer (cottage cheese) is used for curries, pakodas, etc.

Cheese	*Origin*	*Description*
Bel paese	Italy	Mild Italian cheese. Large round and flat, creamy colour, texture soft. Mild sweet flavour, ripens quickly.
Brie	France	Soft texture, colour is reddish, brown crust, yellow paste.
Petite Suisse	Swiss	White colour, creamy texture, prepared from unsalted fresh cream.
Camembert	France	Colour and texture soft and creamy.
Gorgonzola	Italy	Semi-hard blue mould cheese, prepared from renneted cow's milk.
Parmesan	Italy	A hard white cheese. A kitchen cheese. Takes 2 to 3 years to mature.
Emmenthal	Swiss	In colour, it is ivory to light yellow, waxy texture. Has holes approximately of the size of a cherry.
Gruyere	Swisis	As above, but firmer in texture, holes are smaller, the size of a pea. Takes longer to mature than Emmenthal.
Gouda	Dutch (Netherlands)	Dutch cheese pale yellow in colour. Waxy texture. Flat.
Edam	Dutch	Dutch cheese red rind, yellow in colour. Texture waxy. Round like a cannon ball with a flat top and bottom.
Roquefort	France	Blue mould cheese similar to Gongonzola. Has a semi-hard texture.
Neufchatel	France	A soft creamy cheese.
Limburger	Belgium	Soft highly flavoured cheese.
Brick	America	Very firm to fairly soft and slastic texture. Mild to sharp flavour.

Cheese	*Origin*	*Description*
Port du salat	France	Semi-hard, rubbery texture.
English Cheese		
Cheddar	England	Colour yellow, texture close. Clean and mellow, nutty flavour, many uses in the kitchen.
Cheshire	England	White or orange in colour. Texture crumbly. A strong flavoured cheese.
Stilton	England	Finest English table cheese. It is a hard blue mould. Texture flaky. It is usually spooned out for service.
Lancashire	England	It is of a mild flavour, crumbly texture and when fully mature, can be spread like butter.
Wensleydale	Yorkshire	It has a flaky texture and is pale in colour. It has a "honeyed" aftertaste.
Caerphilly	Wales	Creamy white colour, semi-smooth texture and delicate flavour.
Derby	England	Close textured, pale honey colour. Mild when young but when fully matured, has much fuller flavour. Sometimes flavoured with sage.
Double Gloucester	England	Corn coloured, velvety textured. Similar to Cheddar in taste.
Dorset Blue	England	Crumbly texture, strong rich flavour.
Leicestershire	England	Russet coloured, slightly flaky texture and a mild flavour.

Sandwiches

Sandwiches have always been a tradition where catering has been offered to the people. Many ideas have been put forward as to their origin, the most feasible one being that the name came from the Earl of Sandwich, a compulsive gambler.

To save time from the table, he asked for a piece of meat to be placed between two slices of bread which he ate whilst playing. Sandwiches are one of the most varied types of food prepared. They are usually one or two slices of bread, buttered with a filling of meat, chicken, fish, cheese or vegetable, etc.

Bread may be toasted or fresh, a variety of shapes with an almost endless assortment of fillings. The services of sandwiches varies with the purpose for which it is used. Snackbar sandwiches are substantial and a meal unto themselves. With rapid industrialisation and the speed in tempo of work, the use of sandwich as a meal in itself is becoming popular. It is served for tea parties, barbecues, supper or with coffee, used for picnic baskets.

Sometimes sandwiches are taken as a full meal with fruit and milk. Accompaniments of sandwiches are wafers, chiffonade (shredded) lettuce, watercress, juliennes of cabbage or carrots. The different kinds of bread that are used for the preparation of sandwiches are: White bread, brown bread, French loaf and bread rolls, or boat shaped bread.

Approximate quantities for average portions:

Bread: Allow 2 slices of bread per serving. The thickness and number of slices will vary in different catering establishments.

Buns: One sandwich bun per serving.

Butter: Softened 500 gms. butter will spread 96 bread slices if one teaspoonful is used per slice.

Bread Slices: 2 Ib. sandwich loaf white bread cuts 28 $^1/_2$) slices or 36 ($^3/_8$") slices with end crust. Approximate quantities for average portions for various fillings of sandwich are:

Cheese Sliced: 480 gms. average 16 slices (approx. 30 gms. each).

Chicken: 500 gms. average 12 slices or 16 slices.

Peanut Butter: 455 gms. average 13/4 cups.

Tomatoes: 455 gms. average 3-4 medium tomatoes.

Other fillings used are cucumber slices, tomato slices, tuna fish, hard-boiled eggs, meat, olives, etc.

Cutting of Sandwiches: It is done in a number of ways, the type of filling and the method of serving the sandwiches determine the way to cut it.

Method of cutting filled layers into sandwiches, if it is to be prepared in bulk. Important points to be observed for the preparation of sandwiches:

(1) Use stale bread (one day old), because it cuts better and finer.
(2) The bread should have close fine texture.
(3) Thinly sliced bread appeals more than thick slices.
(4) Crust should be removed neatly.
(5) Use a sharp knife preferably with a saw-edge and dip occasionally in hot water.
(6) Store bread at 75°F to 85°F (Ideal). Refrigerator temperature accelerates staling, though in moist, warm climate, store in refrigerator to prevent mould growth.
(7) Sandwiches may be stored for 12-24 hours at temperature below 40°F.
(8) Cream fresh butter facilities easy and even spread, as the creamy consistency glides over the slice.
(9) Use only fresh ingredients, e.g. crisp lettuce, watercress, tomato, etc.
(10) The filling should consist of ingredients which do not clash in flavour, e.g. avoid mixing a sharp flavoured cheese with anchovies.
(11) If the filling contains moisture, apply it just before serving, otherwise the sandwiches become soggy and will appear unappetising.
(12) If sandwiches are prepared well in advance, they should be wrapped mostly in a foil or alternatively

in a sterilized damp napkin; this keeps them supple and fresh.

(13) For serving, they may be dressed on a paper doiley surrounded with lettuce leaves, chiffonade. etc.

Savoury Butters, or Spreads

Lemon: 225 gms of butter, 1 teaspoon of lemon juice, and fish paste, used for fist sandwiches.

Onion: 225 gms. of butter. 2 teaspoons onion juice. 1 teaspoon of water--Cold meats and sandwiches.

Olive: 225 gms. butter, $^1/_4$ cup minced olives--for fish or chicken sandwiches.

Sardine: 225 gms. butter 1/3 cup mashed sardines. 2 teaspoons of lemon juice.

Paprika: For watercress or tomato sandwiches.

Mayonnaise: For egg and salmon sandwiches.

French Mustard: Butter and French mustard for cheese and tinned meat sandwiches.

English Mustard: Butter and English mustard for ham and beef sandwiches.

Types of Sandwiches

(1) *Sandwich Cake*--These look like iced layer cakes and for these, round bread loaves are used. They are sliced in two or three round slices, soft icing containing butter, cheese, etc. It is cut into wedges like cakes.

(2) *Loaf--The* loaf bread is taken and sliced lengthwise, filled and iced in the same way as sandwich cake. It is sliced across the bread and served.

(3) *Pinwheel*--The top crust is taken out of the sandwich loaf. It is then sliced lengthwise, and butter and

soft filling spread evenly, which is of a colour to contrast with the bread. Each slice is rolled and wrapped firmly in a foil and chilled till it is set. It is cut into thin slices and served.

(4) *Ribbon*--They are prepared by using two coloured breads. Two slices of brown bread and one of white or vice versa. The sandwiches are made with three slices, and two different coloured fillings are used. They are cut into thin strips and served.

(5) *Double Decker*--Three slices are used, the filling placed on top. The next slice is buttered and placed on top, lettuce, tomato and cucumber placed on top of this slice. The filial slice is placed in position. Pressed, trimmed and held together with a plastic cocktail stick with stuffed olive or cherry.

(6) *Club*--"fills sandwich is usually served hot and consists of two buttered slices of toast. One slice is the base on which sliced cooked chicken, egg, lettuce, tomato, grilled bacon, and mayonnaise or ham is placed. The other slice of toast is placed on top, pressed, trimmed and decorated. Served between the folds of a paper napkin,

(7) *Bookmaker*--This is buttered roll or French bread spread with French mustard and a grilled steak in between.

(8) *Broadway*--Two layers of toast with smoked salmon, shredded lettuce, sliced hard-boiled egg, served with mayonnaise.

(9) *Conventional Closed Sandwich*--Two slices of bread with any fillings. Cut into two without cutting crust.

(10) *Tea Sandwiches*--Thin slices of bread are used with any fillings cut into triangles and crust removed.

(11) *Buffet Sandwich*--Smaller than conventional, triangles, fingers or squares and fancy shapes.

(12) *French Sandwich*-- Consists of thin slices cut from French loaf and fillings added between two slices.

(13) *Grilled Sandwich*--Two slices of bread with fillings, grilled and then served. They are hot sandwiches.

Open or Smorresbrod: This famous sandwich originated in Denmark and it is an open sandwich having a fancy shape, it is easy to make, appealing to the eye, delicious to eat, and the cost is low, as left-overs can be used.

The choice of filling and decoration is endless. The arrangement on the serving dish is most vital. Doyleys are used and the sandwiches are arranged so that the sandwich's colour scheme and filling blend well.

Examples of Smorresbrod

Capenhagener: Danish blue cheese, with raw egg yolk in raw onion ring.

Louisiana: Danish ham, Camembert cheese, lettuce, glace cheery.

Harlequin: Danish luncheon meat, with potato salad, watercress and radish.

Sunshine Salami: Sliced salami, with raw egg yolk in raw onion ring.

Tivoli Tongue: Sliced tongue, ring green pepper, horseradish, cream, tomato.

Guardsman: Brisket of beef, horseradish, cream, onion, tomato and parsley.

Roskilde Special: Danish ham, topped with chicken mayonnaise, tomato and cucumber.

Kronberg Salad: Ham, topped with Russian Salad and apple, sliced onion.

Frinkadeller: Danish meat balls, red cabbage and fanned gherkin.

Roulette: Cod's roe, Russian Salad, tomato, cucumber, gherkin.

Bacon and Apple: Savoury apple puree, two rashers of cold bacon, lettuce and tomato.

Captain's Breakfast: Pickled herring, slice of egg, capers, on lettuce.

11

General Guidelines

We need all the technical training we can lay hold of. But let us not mistake qualification for ability; it is the use of a tool rather than the possession of it which gets a job done.

—*Vincent Taylor*

1. Disposable plastic gloves are good for clean, safe food preparation.
2. In preparing lettuce for salad cups, remove core by firmly hitting stem end of head flat against counter top. Let lukewarm water run through the lettuce to loosen and separate the leaves. Chill until ready to use.
3. Carrots should not be rubbed with steel wool to remove outer skin because pieces of steel may come off in food and be eaten.
4. An egg slicer may be used to cut cooked carrots, potatoes, hardboard eggs, bananas and other soft foods.
5. To remove the skin from tomatoes, dip them into boiling water until the skin slips off. Cool immediately by placing them in cold water. Place tomatoes in the refrigerator until ready to use.
6. A pair of scissors in the kitchen will come in handy for cutting such foods as meats for salads, and all types of vegetables for salads.

7. Drain canned berries and fruits. Thicken and cook the juice before mixing in the fruit. This keeps the fruit in whole pieces and crisp looking.
8. Do not add salt to large quantities of milk or cream sauce until the end in order to prevent it from curdling. Use a wire beater to make smooth gravies, sauces and puddings.
9. When preparing sections of oranges or grapefruit, cover them with boiling water and allow them to stand for 5 minutes. Score through the skins lengthwise with a sharp fruit knife. Do this twice to divide the skin into quarters, then pull off the rind. This will pull off the white membrane readily, along with the rind, then fruit can be sectioned. Another method is to cut the rind away in a continuing spiral with a slight sawing motion.
10. When making gelatine for moulded salads or sweets, heat only enough liquid to dissolve the gelatine and sugar. To hasten congealing, use ice water to make up the total amount of liquid; or melt the soaked gelatine over hot water, then add remaining liquid
11. When breaking eggs, have a paper sack ready in which to drop to) shells. This saves handling them again when ready to dispose of she' Break one egg at a time into a wet cup before adding it to other'
12. If you are cutting green beans, cut a whole handful at once on your cutting board after stringing--not 2 or 3 at a time.
13. Wrap meat in foil or plastic before storing in freezer or refrigerator --brown paper or waxed paper softens and sticks--this is especially true of minced meat.
14. Instead of stuffed pork chops--try putting on top of chops with a scoop to save time.
15. Bacon grilled on racks saves draining. Bacon cut in cubes while cold, then fried, saves chopping it.

16. Add dry milk, butter and seasonings to hot boiled potatoes, then mash. Add hot water or potato water, as needed for right consistency.
17. To prevent cheese from drying and shrinking, do not grate it too long ahead of time. After grating, keep it refrigerated.
18. Cut orange in half and then slice it. Sliced orange halves with the peel left on are simple to prepare and easy to eat.
19. When serving, use a cup-sized ladle. Using a ladle of the proper size is faster than dipping out several small spoonfuls.
20. When whipping cream, be sure to chill cream, bowl and beater. Cream will then whip in half the usual length of time.
21. Use cooking containers, such as steam table pans and casseroles, for serving foods when possible. The food will be more attractive, stay warm longer and save dishwashing, time and labour.
22. Grind such foods as cheese and meats. This will save time and give uniform results.
23. Use a food chopper to cut such foods as nuts, raisins and vegetables.
24. Dampen raisins with water and they will go through the food chopper without sticking.
25. Use a food scraper to remove food quickly from container.
26. Rub hands with a little oil before preparing pumpkin and other fuzzy vegetables. Prevents roughness and irritation of hands.
27. Soak dishes containing starch and sugar in cold water; greasy ones in hot soapy water. Saves time in cleaning; also saves changing water.
28. When cutting cakes or ice-cream, dip knife in hot water to make a smoother cut.

29. Save time by using correct tools. Measure in large containers, such as 1 litre instead of 4 single cups. Weighing is still more accurate.
30. Always work with both hands. Example: Put lettuce on two salad plates at a time, placing one leaf on each plate.
31. Use large pans for cooking bacon or meat balls and place them in the oven. This will save time watching the dish.
32. Keep two pans of one food on serving counter, so that the line will not be held up when one pan is empty.
33. Use trays or mobile trolleys to bring supplies from refrigerator or supply room.
34. Cutting boards save time in chopping foods and cleaning is made easier.
35. Arrange sequence of work so there is no break or waste of motion. Example: Preparation of apples for baking. Have washed apples placed on level with worker to the left. He picks up an apple with his left hand, cores it and places it in a pan which is placed conveniently to his right.
36. To make celery rings or crescents, cut a bunch of celery with a French knife, sweep celery into colander placed on a level lower than cutting board and wash under strong stream of water.
37. Mix all dry ingredients or liquid ingredients together before blending them into each other.
38. Grease measuring cup slightly before measuring molasses, corn syrup etc., so that nothing is wasted.
39. After using electric meat grinder, put several slices of bread through grinder. This will aid in cleaning the grinder.
40. Butter sides and bottom of pan in which noodles,

spaghetti, rice or apples are cooked to prevent them from boiling over and sticking to the pan.

Recipe Evaluation: The following points should be checked in evaluating a recipe.

1. Is the recipe clear, concise, accurate, and readable as to:
   (a) Ingredients, type or kind
   (b) Amount in weights or measures
   (c) Instructions for method
   (d) Number of portions
   (e) Serving directions.

   Examples:

   (i) Size of serving
   (ii) Garnish to use
   (iii) Serve hot or cold or with a specific accompaniment.
2. Does it produce a quality product as determined by score card? If not, why?
3. Is the product nutritious?
4. Is it economical in time, energy and material?
5. Does it eliminate, as far as possible, the factor of human error?
6. Is the recipe suited to:
   (a) clientele
   (b) available equipment
   (c) workers
   (d) type of service.
7. Is the per capita cost of the product in line with the selling price

Oven Temperatures: In most recipes in this book,

reference has been given to the oven temperature or the gas setting. This is an approximate guide only. Different makes of ovens vary and it is a fact that even the same make of oven can give slightly different results at the same temperature or setting. If in doubt as to whether the temperature given is exactly right for your particular oven, do at all times refer to your own manufacturer's temperature chart. It is impossible in a general book to be exact for every oven, but you will find that the following are a good average in every case.

	Electricity	*Gas Regulator*	
	^0F	^0C	
Cool oven	225-250	107-121	1-9
Very slow oven	250-275	121-135	½-1
Slow oven	275-300	132-149	1-2
Very moderate oven	300-350	147-177	2-3
Moderate oven	375	190	4
Moderate hot oven	400	204	5
Hot oven	425-450	218-233	6-7
Very hot oven	475-500	246-260	8-9
General Temperature Ranges			
	°F	°C	
Very Hot oven	450-75	232-246	For bread, Pastries, Searing of meat
Grilling heat at source	450 and up	232	Grilling meat. fish bacon
Hot oven	400-425	204-208	
Moderate oven	350-375	177-190	
Shallow fry ing/deep frying and girdling	300-375	149-190	
Slow oven	300-325	149-163	
Steam range	228-250	190-128	
Water boils at sea level	212	100	
Flour and corn-flour thickened	203	95	

Simmering, poaching, stewing, braising, good for dissolving gelatine holding coffee and chocolate	185-195	85-91
Maximum for egg custard, Hollandaise sauce	185	85
Final rinse in dishwashing machine (10 seconds)	180	82
Well cooked meat, medium cooked meat (internal temp.)	160	71
Egg coagulate--yolks 76°C white 74°C	156	69
Thermotainers most food for serving	150	65
Hot fudges and other sauces; rare meat	140	60
Danger range in which food bacteria thrive if in moist non-acid food materials	45-120	7.5-49
Holding most fruits, vegetables and dairy products	40	5
Meat storage (short term)	34-36	0
Water freezer	32	0
Holding ice-cream	8-12	-22.5
Holding frozen foods	0 to -20	-29 to -18

Vegetarian Menu

VEGETARIAN LUNCHEON (3 COURSES)

(1) Lentil soup

Vegetable pie with cheese sauce Creamed spinach

or

Vegetable pulao Palak panir, Raita,

Pickle, Papad

Butterscotch sponge Fruit

flan

(2) Tomato juice

Egg croquettes--Tomato sauce

Salad Creamed potatoes

or

Alu ki tahari
Peas panir curry, Dahi bhalla
Chutney, Pickle, Papad
Basundi
Fruit pie

(3) Cream of spinach soup
Vegetable cutlets--Tomato sauce Salad
or
Peas Pulao
Stuffed capsicum, Pakora kadhi Peanut chutney, Papad
Carrot kheer Apricot sponge flan

NON-VEGETARIAN LUNCHEON (3 COURSES)

(1) Mulligatawny soup
Mutton cutlets reform Buttered spinach, Macaire potatoes
or
Vegetable pulao
Mutton khorma, Fried beans, Raita
Pickle, Papad
- -
Coffee mousse
Fruit salad
- -
Coffee

(2) Cream of vegetable soup
- -
Mutton Andalouse
Scalloped potatoes
or
Moghlai biryani, Onion raita
Pickle, Papad
- -
Phirnee
Blancmange
- -
Coffee

(3) Melon
- -

Meat and potato cutlets
Cabbage salad, Sauteed carrots
or
Tomato pulao,
Shami kababs, Dal
Green chutney, Papad
- -
Caramel custard
Basundi
- -
Coffee

VEGETARIAN LUNCHEON (4 COURSES)

(1) Melon cocktail
Cauliflower and gratin, Salad
- -
Lentil rissoles, Tartare sauce
Creamed potatoes, French beans
- -
Methi bhaji pulao
Aloo dum, Masala dal
Chutney, Pickle, Papad
- -
Jalebis
Tangerine chartreuse
Coffee

(2) Minestrone soup
- -
Tomato japonaise, Salad
Nut roiast, Curry sauce
Scalloped marrow
French fried potatoes
- -
Potato biryani
Vegetable khorma
Boondi raita
Pickle, Papad, Chutney
- -
Rasgullas

Bavarois of fruit

- -

Coffee

(3) Cream of asparagus soup

Mock salmon steak, Hollandaise sauce

Macaroni cheese, Salad

- -

Cauliflower pulao

Vegetable kofta curry, Tomato cucumber, Moru kulumbu

Pickle, Papad

- -

Barfi

Baked apple

- -

Coffee

NON-VEGETARIAN LUNCHEON (4 COURSES)

(1) Cream of tomato soup

- -

Fish souffle, Shrimp sauce

Tournedos

French beans, Sauteed carrots

- -

Yakhni pulao

Cucumber raita

Pickle, Papad, Chutney

- -

Neopolitan mousse

Gulab jamun

- -

Coffee

(2) Prawn cocktail

Clear julienne soup

- -

Roast loin of pork

Spiced pineapple sauce

Roast potatoes

Baked cabbage

- -

Moong dal khichdi
Mutton jhal faraizi, Boondi raita
Pickle, Papad, Chutney
- -
Marshmallow pudding
Chocolate sauce
Vermicelli payasam
- -
Coffee

(3) Clear soup royale
Lobster thermidor, Salad
- -
Hungarian goulash with caraway dumplings
Parsley potatoes, Green peas
- -
Savoury coconut rice,
Mutton kofta curry, Cauliflower & pea raita
Pickle, Papad, Chutney
- -
Carrot halwa
Baked coconut pudding
Coffee

LOW COST BALANCED MENUS

(1) Boiled rice (parboiled) Sag (green leafy veg.) with dal
Baingan bagger Curd (with reconstituted non-fat milk)

(2) Tamarind rice
Sag
Groundnut chutney
Dal payasam (reconstituted non-fat milk and jaggery)

(3) Khichdi (parboiled rice, dal and green leafy vegetables)
Dhania chutney Tindli raita (with reconstituted non-fat milk)

(4) Tandoori chapatties
Sprouted gram koshambir
Sarson ka sag
Dahi ki chutney
(curd prepared from reconstituted nonfat milk)

(5) Bajra ki roti

Chhole
Bhathu ki bhajee
Gajar ki kheer (carrot kheer using reconstituted non-fat milk and peanuts)

(6) Tandoori chapatties
Doodhi channa
Groundnut chutney
Palak ka raita (use reconstituted non-fat milk)

(7) Boiled rice
Sambar (with non-fat milk powder and M.P.F.)
Cheera thoran
Buttermilk (use reconstituted non-fat milk)
Boiled tapioca

(8) Dry fish and mango curry
Buttermilk (use reconstituted non-fat milk)

(9) Jackfruit (chakka) curry
Small fish curry
Thakara thoran
Buttermilk (use reconstituted non-fat milk)

(10) Boiled rice
Sambar with peanut flour
Green leafy vegetable Curd (use reconstituted non-fat milk)

(11) Pongal
Mixed vegetable curry
Methi bhaji
Curd curry

(12) Subji, bhat (tapioca macaroni and mixed vegetables)
Pulse chutney Curd (use reconstituted non-fat milk)

(13) Purie (50 percent peanut flour)
Moong dal with sag
Red pumpkin bhujjia

(14) Ragi chapatties

Dal with sag
Onion cuchumber
Curd (use reconstituted non-fat milk)

(15) Ragi uppuma, Sukhi dal
Dahi ka chutney, (use reconstituted non-fat milk)
Palak ka sag

(16) Vegetable uppuma
Onion & green chilli raita
(reconstituted non-fat milk)
Amaranth sag

(17) Bhakri
Mixed dals with sag
Cucumber cuchumber
Dahi (reconstituted non-fat milk)

(18) Boiled rice
Dal with sag
Grilled dried Bombay duck

(19) Boiled rice
Egg and radish curry
Arvi (colocasia)
Amla chutney
Missi roti

(20) Gonglu bhaji (turnip leaves)
Urad dal
Dahi (reconstituted non-fat milk)

(21) Maki di roti, Sarson ka sag
Mixed dais (Channa and urad)
Lassi (reconstituted non-fat milk)

(22) Boiled rice
Varan (green cholai or methi with tur dal)
Curd (use reconstituted non-fat milk)
Groundnut chutney

(23) Chapatties (25 per cent peanut flour)
Palak and onion bhaji
Channa dal chutney

Dahi (reconstituted non-fat milk)
Besan roti

(24) Dahi (reconstituted non-fat milk)
Onion cuchumber
Cholai bhaji

Weight and Volume

		Approximately
Apple (medium)	1	150 gm.
(large)	1	200 gm.
Baking powder (level)	1tsp.	4 gm.
Baking powder (rounded)	1	6gm.
Banana (ripe)	1	125 gm.
(raw)	1	160 gm.
Beetroot	1	150-160 gm
Besan (level)	1dsp.	
" (heaped)	1"	8 gm.
" (level)	1tsp.	13 gm.
" (heaped)	1" cup	10 gm.
Bitter gourd (medium)	1	60 gm.
" " (large)	1	105 gm.
Brinjal (medium size)	1	100 gm.
Buttermilk	1 cup	160 ml.
Cabbage (medium)	1	485 gm.
Capsicum (medium)	1	60 gm.
Capsicum (large)	1	80 gm.
Carrot (medium)	1	65 gm.
" (large)	1	115 gm.
Cauliflower (medium)	1	400 gm.
Celery-	1 bunch	115 gm.
Chicken (standard)	1	1Kg
" (large)	1	1.2 Kg.
" (medium)	1	750 gm.
" (small)	1	500 gm.
Chauli bhaji	1 bunch	140 gm.
Chukka bhaji	1 banch	45 gm.
Cinnamon	5 c.. (2") pice	2 gm.
Coconut (grated)	1	215 gm.

Cocum	3-4	5 gm.
Colocasia leaves	1 bunch	40 gm.
Coriander leaves	1 bunch	100 gm.
Coriander powder (level)	1/4 tsp.	1 gm.
" " "	1/2	2 gm.
" " "	1 tsp.	3 gm.
" " "	1 tsp.	5 gm.
Coriander seeds (level)	1 tsp.	2 gm.
(rounded)	1tsp	3 gm.
(level)	1tbsp.	8 gm.
(rounded)	1tbsp	12 gm.
Cumin powder (level)	1/4 tsp	1 gm..
" " "	1/2 "	2 gm.
Cumin powder (level)	1tsp.	3 gm.
" " "	1tsp.	5 gm.
Cumin seeds (level)	1tsp.	5 gm.
" (rounded)	1tsp.	3 gm.
" (level) Curd	1tsp.	5 gm.
Curd	1 cup	175 ml.
Curry leaves	1 bunch	35 gm.
Dhania jeera powder (level)	1/4 tsp.	1 gm.
" " "	1/2 tsp.	2 gm.
" " "	1tsp.	3 gm.
" " "	1 dsp.	5 gm.
Drumstick	1	30 gm.
Fat	1tbsp.	25 gm.
"	1 cup.	175 gm.
Fenugreek leaves	1 bunch	135 gm.
Flour (level)	1dsp.	8 gm.
" (heaped)	1 dsp	15 gm.
" (level)	1dsp.	10 gm.
" (heaped)	1tbsp	20 gm.
" (heaped)	1cup	100 gm.
Garam masala (level)	1/4 tsp.	1 gm.
" " "	1/2 tsp.	2 gm.
" " "	1 tsp.	3 gm.
" " "	1dsp.	5 gm.
Garlic	2 flales	2 gm.
Garlic	1 pod	20 gm.

Ginger	2.5 cm. (1″) pice	10 gm.
Ginger powder	1 tsp.	2 gm.
Green chilli	1	4 gm.
″ ″ ″	5	10 gm.
Jaggery	1tbsp	15 gm.
″	1 cup	150 gm.
Leeks	1 bunch	70 gm.
Lettuce	1bunch	120 gm.
Mango	1	215 gm.
Mint	1 bunch	115 gm.
Oil	1tsp.	5ml.
″	1dsp	10 ml.
″	1 tbsp	15 ml.
″	1cup	150 ml.
Onion (small)	1	30gm.
″ (medium)	1	65gm.
″ (large)	1	135 gm.
Parsley	1 bunch	30 gm.
Peppercorns	5	1 gm.
Pepper powder (level)	1/4tsp.	1 gm.
″ ″ ″	$^1/_2$ tsp.	2 gm.
″ ″ ″	1tsp.	3 gm.
″ ″ (level)	1dsp.	5 gm.
Pomfret	1	400 gm.
Potato (medium size)	1	120 gm.
(large)	1 bunch	165 gm.
Radish	1	105 gm.
Red chilli	1	1 gm.
	$^1/_4$ stp	4 gm.
Red chilli powder (level)	$^1/_{2stp.}$	1 gm.
	1 tsp.	2 gm.
	1dsp.	3 gm.
Red onion	2 bublbs	6 gm.
″ ″ ″	1 pod	6 gm.
	1 pod	12 gm.
Red radish	1 bunch	70 gm.
Refined flour (level)	1dsp.	10 gm.
″ ″ ″ (heaped)	1 bunch	15 gm.
″ ″ ″ (level)	1dsp	10 gm.

" " " (heaped)	1 dsp.	20 gm.
" " " (heaped)	1tbs.	100 gm.
Rice flour (level)	1dsp.	10 gm.
" (heaped)	1 cup	13 gm.
" (level)	1dsp.	10 gm.
" (heaped)	1 dsp.	15 gm.
" (heaped)	1	1.15 gm.
Salmon (medium)	1	1.5 gm.
Semolina (level)	1	12 gm.
" " " (heaped)	1	16 gm.
" " " (level)	1	14 gm.
" " " (heaped)	1"	135 gm.
" " " (heaped)	1tbsp.	20
Spinach	1 bunch	100 gm.
Sugar	1tbsp.	25 gm.
"	1 cup	175 gm.
Suva bhaji	1 bunch	80 gm.
Tamarind (size of walnut)	20gm.	
(size of lime)	30 gm.	
Tomato (medium)	1	110 gm.
" (large)	1	150 gm.
Turmeric powder (level)	$^1/_4$ tsp.	1 gm.
" " "	$^1/_2$ tsp	2 gm.
" " "	1tsp.	3 gm.
" " "	1dsp.	6 gm.

Rules for Reheating Food

(i) Never recook, only reheat.

(ii) Make use of all scraps of food, e.g., vegetables, sauces, gravy, etc., but be sure that they are absolutely fresh,

(iii) Remove gristle, bone and skin but not fat unless in excess because fat helps to keep the food moist,

(iv) Divide finely so that flavourings and seasonings can penetrate quickly and the food can be reheated quickly to avoid recooking.

(v) Cook any ingredient such as vegetables before adding

it to the food that has to be reheated, e.g., if a cutlet has to be made from roast, the potatoes should be boiled first and then mixed for binding. The short, quick reheating does not allow time for the cooking of raw ingredients,

(vi) Additional moisture is always necessary in the form of sauce or gravy as most of the original moisture will have been removed.

(vii) A binding is essential for croquettes, rissoles, etc., but see that the binding is properly cooked before being added, e.g., as with Panada.

(viii) Cooked meat and fish become insipid when cold. So they must be carefully seasoned and flavoured in reheating. Meat is improved by the addition of ham, tongue, salted meat, onions, tomatoes, mushrooms, curry paste and fresh vegetables, while fish is improved by the addition of lemon juice, parsley and other mixed herbs.

(ix) It is often necessary to coat food to protect it from direct heat and, therefore, merely reheat: e.g.

  (a) Covering with potato as in shepherd's pie.

  (b) Coating with egg and bread crumbs.

  (c) Enclosing in pastry as in mutton patties.

  (d) Enclosing in batter as in pan rolls,

(x) Dish daintily and garnish neatly.

(xi) Serve with a good gravy or a sharp sauce as an accompaniment,

(xii) Serve hot. Half-warmed, reheated food is far from appetizing.

Appendix

SOME POPULAR RECIPES OF INDIA

Some of the popular Indian dishes are as follows:

1. Murgh Anarkali
2. Mutton Rogan Josh
3. Murgh Badam Pasanda
4. Chena Murgi (Sweet Dish)
5. Shahi Tukra (Sweet)
6. Gaujas
7. Aloo Dum Banarsi
8. Keema Matar
9. Nargisi Kofta Curry
10. Paneer Pasanda
11. Moti Pulao
12. Shahjahani Pulao
13. Tikka Kabab
14. Tandoori Pomfret
15. Malai Kofta Curry
16. Murgh Noorjahani
17. Paneer Zalfrezi

Here, we will describe prepration of some of these. You may improve the palatability of same by adding your experience.

Murgh Anarkali

Chicken	--	1 No.
Onions	--	¼ kg
Tomatoes	--	55 gms.
Curd (yoghurt)	--	55 gms.
Ghee (clarified butter)	--	55 gms.
Ginger	--	15 gms.
Garlic	--	10 gm.

Cashew-nuts	--	55 gm.
Illaichi (Cardamom) small	--	1 t.spn.
Eggs	--	2 Nos.
Salt	--	to taste
Paneer (Cottage cheese)	--	55 gms.
Mixed dry nuts	--	15 gms.
Green Dhaniya	--	a few green coriander leaves
Garam Masala whole	--	a little
Coriander powder	--	5 gms.
Red chilli powder	--	5 gms.
Kewra essence or Rose-water	--	few drops
Silver leaves	--	2 Nos.

Method: Clean and cut chicken into four pieces and bone out. Flatten them and stuff with stuffing mixture of paneer and chopped dry nuts. Beat egg yolk and fold into beaten white of egg. Coat the chicken with the egg and deep fry till brown.

Brown chopped onions in hot fat. Add garam masala whole and ground ginger and garlic. Cook for few minutes. Add coriander powder and red chilly powder. Cook till the masala leaves the ghee separate. Add beaten up curd and salt. Add fried chicken and cook till the chicken is tender. Add rose-water. Serve hot garnished with cashew-nuts and illaichi, silver leaves and finely chopped green dhaniya on top.

Mutton Rogan Josh

Mutton	--	500 gms.
Fat	--	30 gms.
Onions	--	155 gms.

Ginger	--	15 gms.
Garlic	--	5 gms.
Turmeric	--	It. spn.
Coriander powder	--	15 gms.
Red chilli powder	--	10 gms.
Tomatoes	--	155 gms.
Salt	--	to taste
Garam masala	--	a little
Green coriander leaves	--	¼ Bunch.

Method: Cut mutton into small pieces. Grind masala to a fine paste. Heat fat and brown sliced onions. Add ground masala. Cook till masala leaves ghee separate. Add tomatoes. Cook for few minutes. Add mutton and cook till mutton is tender. Sprinkle garam masala powder and chopped green coriander leaves upon it. Serve hot.

[Masala--Turmeric, Coriander powder, Red chillies, Garam masala, Ginger and Garlic)

Murgh Badam Pasanda

Chicken	--	4 Nos.
Red chillies	--	15 gms
Onion	--	225 gms.
Ghee	--	115 gms.
Ginger	--	60 gms.
Garlic	--	30 gms.
Turmeric	--	55 gms.
Curd	--	55 gms.
Cashew-nuts	--	55 gms.
Almonds	--	55 gms.
Tomatoes	--	100 gms.
Kewra essence or	--	2 t. spn.
Rose-water	--	2Nos.

Eggs	--	2Nos.
Salt	--	to teste
Silver leaves	--	4 Nos. (Optional)
Green Dhaniya	--	1 Bunch
Pepper	--	1 Pinch
Milk	--	15 ml.
Cream	--	30 ml.
Coriander powder	--	10 gms.
Garam masala whole	--	5 gms.

Method: Clean and cut chicken into four pieces and bone out. Marinade with curd, salt and pepper powder. Dip in flour and beaten egg batter, and coat with chopped cashew-nuts and almonds. Deep fry till light brown and remove. Brown minced onions in hot fat. Add ground ginger and garlic and garam masala whole. Cook for a few minutes'. Add red chillies, coriander powder and cook till it leaves the ghee separate. Add chopped tomatoes and beaten up curd, and cook for few a minutes. Add fried chicken. Cook till the chicken is tender. Grind cashew-nuts using milk. Add cashew-nuts to chicken and cook for a few minutes. Add kewra essence (optional). Serve hot in flat dish garnished with silver leaves (optional). Sprinkle blanched 5 almonds and chopped dhaniya on top.

Chena Murgi (Sweet Dish)

Paneer	--	25 gms.
Sugar	--	115 gms.
Kewra essence or Rose-water	--	few drops.

Method: Cut the paneer into small cubes. Make a sugar syrup and clarify it. Reboil and add paneer cubes. Cook till syrup becomes quite thick. Remove it from the fire, and stir continuously till sugar crystallises and becomes cool. Spread kewra essence (optional) and use as required.

Shahi Tukra (Sweet)

Bread slices	--	10 Nos.
Ghee	--	to fry
Sugar	--	100 gms.
Rabri or condensed milk	--	115 gms.
Khoya	--	115 gms.
Milk	--	a little
Silver leaves	--	16 Nos.
Cardamom (small)	--	4 Nos.
Saffron colour	--	few drops
Kewra essence or Rose-water	--	few drops
Pistachio nuts	--	15 gms.

Method: Remove the crust of bread slices and deep fry to a golden brown colour. Heat milk and sugar. Add fried bread slices when it is quite thick. Add condensed milk and all the ingredients except nuts and silver leaves. Cook on slow fire till milk becomes very thick. Arrange the bread slices in plates. Pour rest of the milk mixture over it, cover with silver leaves, sprinkle shredded nuts on top and serve warm.

N.B.: Sugar should be put according to the sweetness required.

Gaujas

Flour	--	115 gms.
Fat	--	30 gms.
Salt	--	a pinch
Cardamom	--	4 Nos.
Sugar	--	115 gms.
Water (syrup)	--	30 ml.
Lime juice	--	few drops

Method: Mix flour and salt and prepare a stiff dough using a little water. Divide into oval sized balls. Roll out into thin poories (flats or rounds) about 3" in diameter. Make slits in the centre taking care that the ends are not separated. Roll from either sides to the centre, fold ends and twist. Deep fry in hot fat, remove and dip in prepared sugar syrup. Sprinkle crushed cardamom over.

N.B.: Powdered sugar may also be sprinkled over fried Gaujas while they are still hot, instead of putting them in syrup.

Aloo Dutn Banarsi

Large potatoes	--	1kg.
Peas	--	115 gms.
Sultanas	--	55 gms.
Ghee	--	115 gms.
Onions	--	115 gms.
Tomatoes	--	115 gms.
Ginger	--	30 gms.
Garlic	--	15 gms.
Red chilli	--	15 gms.
Coriander powder	--	15 gms.
Haldi	--	15 gms.
Garam masala whole	--	6 gms.
Salt	--	to taste
Green dhaniya	--	¼ Bunch
Green chilli	--	5 gms.
Curd	--	55 gms.

Method: Saute little chopped onions. Add boiled peas, salt, green chillies and little red chillies. Cook for a few minutes. Add washed sultanas and use for stuffing. Cut the large potatoes into half and scoop cut inner portion. Stuff with the above stuffing and join both the sides with tooth picks or the batter made of flour, water and egg.

Curry--Brown sliced onions in hot fat. Add garam masala whole. Add ground masala. Cook till masala leaves the ghee separate. Add chopped tomatoes and cook for a few minutes. Add beaten up curd. Add brown potatoes and cook till the potatoes are tender. Sprinkle chopped dhaniya on top. Serve hot.

Keema Matar

Minced mutton	--	½ kg.
Peas	--	¼ kg.
Red chilli	--	115 gms.
Tomatoes	--	10 gms.
Garlic	--	5 gms.
Garam masala powder	--	5 gms.
Ghee	--	115 gms.
Salt	--	to taste
Coriander powder	--	5 gms.
Haldi	--	5 gms.
Garam masala whole	--	a little
Onions	--	115 gms.

Method: Prepare ordinary gravy out of onions, tomatoes, ginger, garlic, coriander powder and haldi. (In hot oil add all the ingredients and fry.)

Add keema and cook for 20 minutes. Sprinkle water as and when required. When tender serve hot.

Nargisi Kofta Curry

Meat minced	--	500 gms.
Eggs boiled	--	4 Nos.
Egg	--	1 No.
Garam masala powder	--	a pinch
Green chillies	--	1 No.
Salt	--	to taste

Fat	--	to fry
Tomatoes	--	500 gms.
Turmeric	--	It. spn.
Red chilli powder	--	It. spn.
Garlic	--	It. spn.
Coriander leaves	--	1 sprig

Method: Mince meat finely. Add beaten egg. Add finely chopped green chillies, onion, powdered spices and salt. Dip eggs in seasoned flour. Coat with minced meat mixture and deep fry. Meanwhile heat fat for curry and saute sliced onions. Add garlic, turmeric and red chilli powder and fry. Add pureed tomatoes and salt and simmer. Drop the koftas one by one into the gravy and simmer. Cut the koftas into two halves and let each piece float in the gravy with yolk side up. Garnish with coriander leaves.

Paneer Pasanda

Paneer	--	275 gms.
Salt	--	to taste
Tomatoes	--	275 gms.
Ginger	--	15 gms.
Kashmiri mirch	--	to taste
Flour	--	55 gms.
Green dhaniya	--	a little
Green chillies	--	2 Nos.
Minced dry nuts	--	15 gms.
Garam masala	--	5 gms.
Onions	--	55 gms.
Garlic	--	5 gms.
Orange colour	--	1 t. spn.
Ghee	--	to fry

Method: Prepare 2" x 1½ slices 1 /8" thick of paneer. Mix together minced dry nuts, pounded left over paneer and chopped green chillies. Put this mixture on one piece and cover it with another. Prepare batter with flour, seasoning and orange colour. Coat the prepared paneer with this batter and deep fry. Make gravy with onion, fat paneer and serve hot. .

Paneer: Boil 2 litres of milk, add one lime-juice and tie in a muslin cloth, and let water drive out.

Moti Pulao

Mutton	--	500 gms.
Rice	--	1 kg.
Salt	--	to taste
Seasoning	--	to taste
Onions	--	400 gms.
Saffron	--	a little
Lime-juice	--	few drops
Kewra essence or Rose-water	--	few drops
Sultanas	--	110 gms.
Almonds	--	110 gms.
Silver leaves	--	12 Nos.
Cornflour	--	for binding
Paneer	--	230 gms.
Ghee	--	230 gms.
Pomegranate (Anar)	--	110 gms.

Method: Prepare rice for Kashmiri pulao, serve hot garnished with fried dry nuts, silver leaves, fried small rollers of paneer (with cornflour as binder), pomegranate seeds covered with silver leaves.

Shahjahani Pulao

Rice	--	1 kg.
Minced mutton	--	½ kg.

Egg	--	1 No.
Cashew-nuts	--	50 gms.
White pepper	--	a little
Salt	--	to taste
Curd	--	200 gms
Onions	--	55 gms.
Garam masala whole	--	5 gms.
Lettuce	--	1 bunch
Ghee	--	115 gms.
Silver leaves	--	2 Nos.

Method: Mix minced mutton, ground cashew-nuts, seasoning and egg. Form into small balls and deep dry.

Heat fat, add sliced onions and garam masala whole. When light brown, add washed and soaked rice, salt and beaten up curd. Cook till rice is cooked and the moisture is almost dried. Add mutton balls and cook on dim pressure. Serve hot garnished with silver leaves, brown sliced onions and fried shredded lettuce.

Tikka Kabab

Leg of mutton	--	1 kg.
Curd	--	115 gms.
Black cumin seeds	--	10 gms.
Salt	--	to taste
Orange colour	--	15 ml.
Garam masala powder	--	10 gms.
Lemon	--	4 Nos.
Ground ginger	--	10 gms.
Ground garlic	--	5 gms.
Red chilli	--	5 gms.
Green chilli	--	15 gms.
Onions	--	55 gms.

Method: Cut mutton into small pieces, then marinade in a batter.

Marinade: Mix curd and all ingredients except leg of mutton. Skewer the meat pieces and grill.

Tandoori Pomfret

Pomfret -- 1 kg.
Salt -- to taste
Lemon -- 2 Nos.
Garam masala powder -- 5 gms.
Curry powder -- 5 gms.
Cumin powder -- a little

Method: Make paste with garam masala powder, salt, curry powder and lemon juice. Apply this paste over the clean pomfret and leave aside for few minutes. Cook it in moderate hot tandoor till tender. Serve hot garnished with green chilli, lemon slices and ring onions.

N.B.: Can be grilled if tandoor is not available.

Malai Kofta Curry

Malai/Cream -- 115 gms.
Paneer -- ft kg.
Patatoes -- ½ kg.
Cornflour -- 115 gms.
Butter -- 55 gms.
Sultanas -- 30 gms.
Cashew-nuts -- 115 gms.
Red chilli powder -- 15 gms.
Onions -- 225 gms.
Curd -- 225 gms.
Ghee -- 115 gms.
Whire pepper -- 15 gms.

Salt	--	to taste
Silver leaves	--	2 Nos.
Kewra essence or Rose-water	--	2 t. spn.
Green Dhyania	--	½ Bunch
Ginger	--	35 gms.
Garlic	--	15 gms.
Small Cardamom	--	4 Nos.
Milk	--	4t.spn.

Method: Boil potatoes and mash paneer, and mix. Add salt, white pepper and cornflour and mix well; divide into small balls, stuff with small cubes of frozen butter and one sultanas each. Reshape it and deep fry in hot fat till brown and remove. Saute minced onions in hot fat, add ground ginger, garlic and green cardamom. Cook for few minutes. Add red chillies and white pepper. Add beaten up curd and cook. Add ground cashew-nuts (grind cashew-nuts using milk and salt). Cook for 15 minutes. Add kofta and simmer on fire for 15 minutes. Add cream and boil it. Add kewra essence or rose-water, Serve hot garnished with silver leaves and chopped dhaniya on top.

Murgh Noorjahani

Chicken	--	2 Nos.
Chicken liver	--	12 Nos
Eggs	--	4 Nos.
Cashew-nuts	--	55 gms.
Tomato sauce	--	15 gms.
Mustard powder	--	It. spn.
Ginger	--	60 gms.
Garlic	--	30 gms.
Curd	--	115 gms.
Egg (hard boiled)	--	1 No.

Onions	--	225 gms.
Ghee	--	115 gms.
Tomatoes	--	55 gms.
Red chillies	--	15 gms.
Haldi	--	5 gms.
Kewra essence or Rose-water	--	2 t.sp.
Flour	--	2 t.sp.
Garam masala		
Cinnamon, cloves, cardamom		

Method: Chop chicken liver and saute with a little chopped onions in hot fat. Add tomato sauce and use for making bed for chicken.

Make the mixture of curd, ginger, garlic, mustard powder flour and eggs. Cut chicken into four pieces and bone out and marinate chicken in the above mixture. Deep fry chicken in hot fat till light brown and remove. Heat fat and brown chopped onions. Add garam masala whole, chilli and haldi. Cook till the masala leaves ghee separate, add chopped tomatoes. Cook till the tomatoes are mashed Add beaten up curd and cook. Add fried chicken. Simmer in tender and slow fire. Add ground cashew-nuts and cook till the chicken is tender. Add kewra, serve on bed of cooked chicken liver and garnish with chopped hard boiled egg on top.

Glossary

A la broche: Roasted in front of a fire on a spit or in a Dutch oven.

A la carte: Opposite of table d'hote, meaning that each dish is ordered and priced individually. A

A la frangais: Dishes that are prepared in the French way.

A la: Meaning merely, "in the style of.

A peritif: A cocktail or other drink served before a meal to stimulate the appetite.

Abats: Butcher's supplies such as heads, hearts, liver, kidney, etc.

Aging barrel: A barrel used to age wine or distilled spirits.

Aiguillettes: Thin slices of breast of poultry cut lengthwise.

Akhni or Yakhni: Soup or stock.

Al Dente: An Italian term meaning "to the tooth." Pasta is cooked tender, but still firm.

Alcohol: Generally refers to ethanol, a chemical compound found in alcoholic beverages. It is also commonly used to refer to alcoholic beverages in general.

Alu: Potato.

Amphora: A type of ceramic vase, used for transporting and storing wine in ancient times.

Amriti: A sweetmeat.

Angel's share: The portion of a wine in an ageing barrel that is lost to evaporation.

Anthocyanin: Phenolic pigments that give red wine its colour.

Appellation: A geographical based term to identify where the grapes for a wine were grown.

Appetizer: A tidbit served before a meal or as the first course. A drink such as sherry, cocktail, etc.

Aroma: The smell of a wine. The term is generally used for younger wines, with the term *Bouquet* used for aged wines.

Aspic: A transparent savoury jelly, generally made of seasoned meat stock. It is used to garnish meat or fish, or to make moulds of meat, fish or game.

Atta: Wholemeal flour.

Au bleu: Applied to fish cooked in fish stock with wine added. Au four Baked in the oven.

Au gratin: Any dish covered with sauce, breadwinners or cheese and afterwards baked or grilled. The food is served from the dish in which it is cooked.

Au maigre: Opposite of au gras. Applied to meatless or Lenten dishes.

Au naturel: Simply-cooked food, or food served raw, such as oysters.

Baffad: A curry with meat and radish from Goa.

Baghar: Tempering done after the dish has been prepared. Onions and a few spices or herbs are fried in a spoonful of fat and added to the cooked dish to improve flavour.

Bain-Marie: A large shallow pan to hold water in which several small sauce-pans can be heated without their contents boiling, an alternative to the double saucepan, but often used to keep cooked food warm.

Bake: (i) Cooking in an oven. (ii) To cook by dry heat, usually in an oven.

Baked hilsa: An oily fish considered a delicacy. Smoked slightly--all bones removed and baked with anchovy essence, tomato sauce, salad oil, mustard, etc.

Balushai: A round ball made of dough, slightly flattened at the centre, fried and dipped in sugar syrup.

Barbecue: Meat basted with a highly- seasoned sauce, e.g., lamb basted with hot red-currant sauce.

Barfi : A fudge-like sweet.

Barquette: A boat-shaped pastry tartlet, filled with chicken, vegetables, oyster, fish mayonnaise, etc.

Barrel: A hollow cylindrical container, traditionally made of wood staves, used for fermenting and aging wine. Sometimes called a *cask*.

Barrique: The French name for a 225 litre Bordeaux style barrel.

Baste: The process of spooning or pouring pan juices over food while it cooks. This promotes moistness and a browned surface.

Basundi: Milk sweet.

Batter: A mixture of flour and liquid such as milk, egg, etc., of such consistency that it can be beaten or stirred. Used to coat foods for frying, or as pancakes, etc. May be sweet or savoury.

Baumé: A measure of the sugar concentration in the juice or wine.

Bearnaise: A rich sauce, resembling a Hollandaise sauce, made from butter, egg yolk and vinegar, and flavoured with peppercorns, shallots, chervil and tarragon

Beat: To mix air with food by vigorous motion; also used to make a mixture smooth and free from lumps. A wooden spoon is best for beating thick mixtures; an egg-whisk for thin ones.

Bechamel: a rich white sauce.

Beeswing: A light sediment, chiefly mucilage, found in Port.

Beignets: Pancake batter fried in deep fat or fritters of different kinds.

Bentonite: A type of clay used in wine *clarification*.

Beurre manie: Equal quantities of flour and butter used for thickening sauces.

Beurre: noir Browned butter.

Bhaji: Another term for vegetable preparations.

Bhajjia: Slices of vegetables dipped in gram flour batter and fried crisp.

Bhaturas: Slightly leavened bread baked on a grid and then fried. Generally served with 'Chhole', a dish prepared out of whole chickpeas.

Bhel puri: Crisp fried thin rounds of dough mixed with puffed rice, fried lentil, chopped onion, herbs and chutney.

Bhindi: Okra/gumbo, an Indian vegetable known as ladies fingers.

Bhujjia: A North Indian vegetable preparation.

Bhurta: Vegetable boiled or roasted in charcoal, peeled, mashed and sauteed with a little chopped onion and chillies.

Biodynamic wine: Wines produced by the principles of biodynamic agriculture.

Biryani: A rich preparation of rice. Parboiled rice is put in layers with a rich meat or vegetable curry and the whole is baked till cooked.

Bisque: A rich, thick, cream soup, usually made from shellfish, e.g. lobster.

Black Papper: The berry of the pepper vine dried with the skin on.

Blanc de Blancs: A white wine made from white grapes.

Blanc de Noir: A white wine made from red grapes.

Blanch: Literally to whiten, but used here to mean dipping food into boiling water for a few moments, and then into cold, to remove skin. To blanch almonds, boiling water is poured on them to whiten them and to remove the skins.

Blanch: Refers to partially cooking an ingredient, usually vegetables, by placing them in boiling water for a few minutes.

Blend: (i) Combining ingredients until smooth and uniform in texture, flavor, and color. (ii) To combine two or more ingredients.

Blending: The mixing of two or more different parcels of wine together by winemakers to produce a consistent finished wine that is ready for bottling. Laws generally dictate what wines can be blended together, and what is subsequently printed on the wine label.

Blind tasting: Tasting and evaluating wine without knowing what it is.

Bodega: A Spanish wine cellar. Also refers to a seller of alcoholic beverage.

Boil: Heating liquid, which causes a constant production of bubbles that rise and break the surface.

Bombay Duck: A small, phosphorescent, gelatinous fish, abundant at the surface of the salt waters of Bengal and Bombay. It is prepared fresh or dried. When dried, it is toasted crisp and served.

Bombe: glacee A mould lined with one kind of ice cream and then filled with ice cream of a different flavouring.

Bonbon: Sugar confectionery -- sweets.

Bonda: Mashed potatoes, seasoned and formed into balls, dipped in batter and deep fried -- a South Indian snack.

Bottle shock: Also known as bottle-sickness, a temporary condition of wine characterized by muted or disjointed fruit flavors. It often occurs immediately after bottling or when wines (usually fragile wines) are shaken in travel. After several days the condition usually disappears.

Bottle variation: The degree to which bottled wine of the same style and vintage can vary.

Bouchees: Small patties of light pastry sufficient for one mouthful.

Bouillon: Unqualified broth or stock made from fresh meat.

Bouquet garni: A small bunch of mixed herbs used for flovouring soups, stews, etc. Ideally, this should consist of a sprig each of parsley, thyme, basil, marjoram, with a bay leaf and a strip of lemon peel, tied together with cotton. It must always be removed before the dish is served. When fresh herbs are unobtainable, dried herbs tied in a piece of muslin may be used instead.

Bouquet: A tasting term for the complex aromas of an aged wine. The term is generally not applied to young wines.

Box wine: Wine packaged in a bag usually made of flexible plastic and protected by a box, usually made of cardboard. The bag is sealed by a simple plastic tap.

Bradsot: Russian soup containing beetroot.

Braise: Cooking food slowly in a small amount of liquid in a tightly covered pan.

Brettanomyces: A wine spoilage yeast that produces taints in wine commonly described as *barnyard* or *band-aids*.

Bright: Describes a wine that has high clarity, very low levels of suspended solids.

Brinjal: Eggplant or aubergine. Also called baingan.

Brioche: Very light French rolls.

Brix: A measurement of the dissolved sucrose level in a wine.

Broil to cook: By exposing food directly to heat; used synonymously with "grill".

Broil: Cooking food using an overhead source of heat.

Browning: A substitute added to stews and gravies to darken them.

Brunoise fashion: Cut into small dice.

Brush: A thin, even coating of beaten egg or milk applied to pies, buns, etc., immediately before they are put in the oven. It gives the pastry or bun a glossy appearance and helps it to brown more quickly and deeply. Pastry brushes are sold for this purpose. Dried egg is especially useful for brushing, as a very small quantity can be mixed instead of beating a whole egg.

Brut: A French term for a very dry champagne or sparkling wine. Drier than extra dry.

Bung: A stopper used to seal a bottle or barrel. Commonly used term for corks.

Butt: An old English unit of wine casks, equivalent to about 477 litres (126 US gallons/105 imperial gallons).

California cult wines: Certain California wines for which consumers and others pay higher prices than those of Bordeaux's First Growths (Premiers Crus).

Canapes: Small pieces of toast, fried bread or pastry on which light savouries are served.

Canneloni: Small pasta rolls filled with mince (Italian).

Capon: A castrated male chicken. It grows large and has tender meat.

Capsule: The plastic or foil that covers the cork and part of the neck of a wine bottle.

Caramel: A substance made by heating sugar until it turns dark brown. Used for coating moulds, flavouring dishes, etc.

Caramelize: To heat sugar until it turns brown.

Carbonic maceration: A winemaking practice of fermenting whole grapes that have not been crushed.

Cardamom: A spice--the fruit of the cardamom tree. The

seeds are found in a pod which is dried. It has a pleasing taste and flavour.

Carmelize: The process of causing sugar or the natural sugars in food to darken to a golden brown and develop a rich flavor by cooking on a constant heat.

Carte do jour: Menu for the day.

Cashewnut: The nut of the cashew fruit. It is found outside the fruit, enclosed in a kidney-shaped shell.

Casserole: a baking dish with a well-fitting lid used for cooking stews, etc. in the oven. Usually made of fireproof earthenware or glass, or of heavy enamel. The food is usually served from the casserole.

Cassolette: A kind of hot hors d'oeuvre moulded to the shape of a small drum.

Caviar: Salted roe of sturgeon or similar fish.

Cepes: A kind of mushroom.

Cereals: Grains such as wheat, including semolina, oats, barley, rye, rice etc.

Champagne flute: A piece of stemware having a long stem with a tall, narow bowl on top.

Chapatti or Phulka: Wheat bread made of unfermented dough, round in shape and paper thin. Resembles a tortilla.

Chaptalization: A winemaking process where sugar is added to the must to increase the alcohol content in the fermented wine. This is often done when grapes have not ripened adequately.

Chardonnay: A type of wine, one of the "noble" white varietals.

Charlotte russe: Generally a mould lined with sponge cake or sponge fingers and filled with a mixture of cream and fruit, and jelly.

Charlotte: A sweet made of alternate layers of fruit and either bread crumbs or slices of bread and butter.

Charmat process: The Charmat or bulk process is a method where sparkling wines receive their secondary fermentation in large tanks, rather than individual bottles as seen in Méthode champenoise.

Charoli: An Indian nut.

Chartreuse: Mould of fruit, jelly or savoury mixture.

Château: Generally a winery in Bordeaux, although the term is sometimes used for wineries in other parts of the world, such as the Barossa Valley.

Chateaubriand: Head of the fillet of beef.

Chaud-froid: A cold sauce used for coating meat, game or fish.

Cheera: Amaranth leaves.

Cheese fondue: Grated cheese melted in white wine, seasoned with pepper and flavoured at the last minute with a little kirsch.

Chiffonier: Mince of sorrel, lettuce, etc. Wash the leaves, remove veins, roll and cut into fine strips with a sharp knife.

Chikkoo: A brown Indian fruit--a cross between a fig and a russet apple. Also known as sapota.

Chilopoda: A kind of small sausage.

Chinos: A conical strainer.

Chiwda: A mixture of nuts, fried pressed rice, fried lentils, gram preparations and spices.

Chop: Cutting foods with a knife into smaller pieces.

Choux: A kind of pastry used for such things as cream buns and eclairs.

Chowder: An American soup made with pickled pork, shellfish, fish, potatoes, and other vegetables.

Cisel: Finely chopped vegetables.

Claret: British name for Bordeaux wine. Is also a semi-generic term for a red wine in similar style to that of Bordeaux.

Clarification: A winemaking process involving the *fining* and *filtration* of wine to remove suspended solids and reduce turbidity.

Cleanskin: Wine bottled without a label.

Coat: To cover with a thin layer.

Cocum: A sour fruit dried and used chiefly with fish. Acid in taste.

Cold Duck: A mixture of red and white sparkling wine that has a high sugar content.

Cold stabilization: A winemaking process where wine is chilled to near freezing temperatures for several weeks to encourage the precipitation of tartrate crystals.

Concass: To chop roughly, e.g., concassed tomatoes.

Condiments: Spices and seasonings.

Confiture: Jam or fruit preserves.

Congalaise: Garnish of poached oysters and shrimps cohered with Normande sauce.

Consistency: The thickness or texture of mixture, such as a cake or batter mixture.

Consomme: A light-coloured clear soup.

Coriander: A spice, the seed of a small herb.

Cork taint: A type of wine fault describing undesirable aromas and flavours in wine often attributed to mould growth on chlorine bleached corks.

Cork: A wine bottle stopper made from the tissue of the cork oak tree.

Corked: A tasting term for a wine that has cork taint.

Corkscrew: A tool, comprising a pointed metallic helix attached to a handle, for drawing Corks from bottles.

Cotelettes: Cutlets.

Coupe: A cream or water ice served with fruit.

Court-bouillion: A well-flavoured cooking liquor for fish.

Crackling: Semi-sparkling wine; slightly effervescent. Also called *frizzante*.

Crazy (a la): Dishes containing carrots.

Cream fat: To beat fat with a wooden spoon until it is light and fluffy.

Crémant: French sparkling wine not made in Champagne region.

Croissants: French rolls, crisp and light.

Croquettes: Leftover meat, fish, poultry or game, finely minced and rolled into small sausage shapes. These are coated with egg and bread crumbs and fried a golden brown.

Croutes: Ovals or rounds of fried bread.

Croutons: Bread cut in small dice or fancy shapes and fried or toasted. Used as a garnish for serving with soup.

Crust: Sediment, generally potassium bitartrate, that adheres to the inside of a wine bottle.

Cuisse: Leg, e.g., cuisses de volailles, cuisses de granules.

Cult wines: Wines for which committed buyers will pay large sums of money because of their desirbility and rarity.

Cumin: A spice resembling caraway seeds.

Curry powder: A mixture of various spices such as coriander, chillies, turmeric, cumin etc., roasted and powdered together.

Custard apple: Resembles a leaf artichoke. A luscious custard-like fruit with hard black seeds inside.

Custard: A cooked or baked mixture made of milk and eggs; it may be sweetened for dessert, or it may be flavoured with cheese, fish, etc., as an entree.

Cut and fold: To mix flour very gently into a mixture.

Cut In/ Working a solid fat such as butter or shortening into dry ingredients.

Cutlet: A small piece of meat cut usually from rib of veal or pork, mutton or lamb, usually grilled or fried.

Cuvee: The pressing, or a blending of several wines.

Dahi Bhalla: Ground lentil, made into fried balls and then steeped in seasoned beaten yoghurt. Served with pulao or as a snack.

Dahi: Yoghurt or curd.

Dal Lentil: There is a large variety of lentils in India.

Dariole: A small cup-shaped mould.

Darne: The middle slice of a fish.

Dash: 1/16 teaspoon measure. (Use 1/2 of a 1/8 teaspoon measure.)

Decanting: The process of pouring wine from its bottle into a decanter to separate the sediment from the wine.

Deglaze: Adding stock, wine or water to the pan after meat has been cooked and removed. When the liquid is added, it loosens the flavor from the pan to make a sauce for the meat.

Dégorgement: The disgorging or removal of sediment from bottles that results from secondary fermentation.

Demi-sec: Moderately sweet to medium sweet sparkling wines.

Dessert wine: Very sweet, high alcohol wines.

Devilled: The same as 'a la doable', that is, any highly seasoned dish. Often this seasoning is done with some form of curry powder.

Dhansak: Literally meaning "rice & vegetables", a Parsi speciality. It is plain fried rice served with a curry rightly called "wide mouthed" as it contains an innumerable variety of ingredients--meat, lentils, vegetables, leafy vegetables, nuts and a variety of spices.

Dice: To cut into small even cubes. Cut food first into

slices 1/2 in thick, then into strips the same width and finally hold strips together and cut into cubes. Use a chopping board and a sharp knife.

Dieppoise: A garnish for fish consisting of shrimp's tails, mussels and mushrooms.

Dissolve: Stirring a solid food into a liquid to form a mixture in which the solid food does not remain.

Do pays: Literally "twice onion". Ground and fried onion added to meat along with other spices. Hence Do (twice) and Pays (onion). It is not very hot and could form a tasty accompaniment to parathas and chapatties.

Doodh path: Sweetmeat made of rice and milk.

Dosa: A type of savoury pancake made with fermented batter of ground lentil and rice. Cooked on a griddle. A speciality of South India.

Dot: To put small bits of butter, cheese, cream, etc., over the surface of a dish. Dough A mixture of a liquid, flour, etc., kneaded together into a stiff paste or roll.

Doux: The French word for sweet. Usually refers to the sweetest category of sparkling wines.

Dredge: Coating food with flour, corn meal or sugar before of after cooking.

Drip dickey: A wine accessory that slips over the neck of a wine bottle and absorbs any drips that may run down the bottle after pouring - preventing stains to table cloths, counter tops or other surfaces.

Drumstick: A long type of bean with hard fibrous covering, resembling a drumstick.

Dry: Wines with zero or very low levels of residual sugar. The opposite of sweet.

Dum phukta: A dry rich meat dish prepared by cooking meat in pressurised steam.

Dura: A process of cooking with heat from above as well as below.

Dust: To sprinkle lightly with fine sugar, dried milk or flour.

Duxelle: A mixture of parsley, mushrooms and shallots used for flavouring sauces and purees or as a forcemeat.

Ecarlate: Salted meat is said to be a l'ecarlate when it is covered with a coat of red jelly.

Eclair: Choux pastry filled with custard or cream. Can be coated with chocolate.

Eiswein: German for ice wine, a dessert wine made from frozen grapes.

Emince: Finely sliced or shredded.

Emulsion: Suspension of two liquid ingredients that do not dissolve into each other.

en Tirage: French for "*in pulling*", refers to the period of time in which bottled sparkling wine is rested in contact with lees generated during secondary fermentation. Part of the Méthode Champenoise process.

Enology: American English spelling of oenology, the study of wine.

Entree: A dish served in the first part of a dinner, usually a made-up dish with sauce. Or it can be the main dish of a less formal meal.

Entremets: Hot or cold sweets.

Escalopes : Thin slices of meat dipped in egg and bread crumbs and then fried.

Espagnole: A rich brown sauce.

Estouffade: Brown stock.

Extra dry: A champagne or sparkling wine with a small amount of residual sugar (slightly sweet). Not as dry as *Brut*.

Failles: Potato straws.

Farci: Any kind of stuffing.

Farm winery: A United States winery license allowing farms to produce and sell wine on-site.

Fennel: A fragrant herb used for flavouring sauces (saunf).

Fermentation: The conversion of grape sugars to alcohol by yeast.

Fiasco: The straw-covered flask historically associaed with Chianti.

Fighting varietal: A term that originated in California during the mid 1980s to refer to any inexpensive cork-finished varietal wine in a 1.5 liter bottle.

Fillet: To bone fish, etc., and cut into fillets.

Fillets: Fish with bone removed. Undercuts of veal or beef. Slices from the breast of a bird.

Fining: A *clarification* process where flocculants, such as bentonite or egg white, are added to the wine to remove suspended solids.

Finish: A tasting term for the lingering aftertaste after a wine has been swallowed.

Flagon: A glass bottle that holds two litres of (usually inexpensive) table wine.

Flan: A pastry case made in a flat tin and afterwards filled with a sweet or savoury mixture.

Flor: The yeast responsible for the character of dry Sherries.

Flute: A long crisp roll of bread (French) used to garnish soups or to serve with soup. Foie gras Liver of a fat goose.

Fold: A method of gently mixing ingredients. Use a rubber spatula to cut down through the mixture, move across the bottom of the bowl, and come back up, folding some of the mixture from the bottom close to the surface.

Fondant: Sugar boiled to 112°C (234°F) and then beaten

to a "fudge-like" smoothness. It can be used as an icing on any kind of sweet.

Fool: A summer sweet made with sieved fruit and whipped into a frothy mixture and served with cream or custard. Forcemeat Savoury stuffing.

Fortified wine: Wine to which alcohol has been added, generally to increase the concentration to a high enough level to prevent fermentation.

Foxy: A tasting term for the musty odor and flavor of wines made from *Vitis labrusca* grapes native to North America.

Frangipane: Confectioner's custard. Frappe Sweetened fruit juice, half frozen.

Free run: Juice obtained from grapes that have not been pressed.

Fricadelles: Braised game or meat in very small pieces.

Fricassee: A stew generally made with chicken, veal, rabbit or lamb thickened with white sauce to which milk has been added.

Fritters: Fruit, meat, vegetables or fish coated with batter and fried, usually in deep fat.

Frosting: A cooked or uncooked sugar icing used to cover and decorate cakes, etc.

Fry: (i) Fast browning and cooking food in varying amounts of fat, most often at a high heat. (ii) To cook food in very hot fat in an open frying-pan. Furnet A kind of essence extracted from fish, game, etc.

Galantine: Cooked meat that has been boned, pressed into a mould with jelly and served cold.

Galette: A large quoit made from puff pastry or short pastry.

Garam masala: A mixture of cloves, cinnamon, cardamom, pepper and cumin.

Garnish: Adding extra appeal to a finished dish.

Gateau: This can be a cake, an ice cream, or anything made in the shape of a cake and lavishly decorated.

Gaujas: A fancy sweetmeat made of dough fried and dipped in sugar syrup.

Gelatine: A product made from refined cows' hoofs. Used for jellies. Sold either granulated or in sheets. The strength varies with each brand. Allow 1/2 oz. to 1/3 oz. to each pint of liquid.

Genoese: A rich butter sponge used as a base for gateaux, and as a base for savoury fingers, etc.

Ghee: Clarified butter or vegetable shortening processed to resemble clarified butter.

Giblets: The heart, liver and gizzard of poultry used to make gravies, soups and pies.

Gild, gilding: (i) To cover an object with beaten eggs by means of a brush, (ii) To give a golden sheen to objects by means of heat.

Glace: Ice, ice-cream, icing, to make smooth or glossy with icing or jelly. Also crystallized or frozen foods.

Glaze: Meat glaze is made by reducing (by boiling) stock or gravy to the consistency of jelly. It is used for improving the appearance of cold meats, etc. Sugar and water glaze is brushed over fruit pies, buns, etc., when they are taken from the oven, to improve their surface appearance. Egg-and-water glaze is brushed on to savoury pies, buns, etc., before they are put in the oven.

Globalization of wine: Refers to the increasingly international nature of the wine industry, including vineyard management practices, winemaking techniques, wine styles, and wine marketing.

Goulash: Hungarian meat stew, flavoured with paprika.

Grape juice: The free-run or pressed juice from grapes. Unfermented grape juice is known as "must."

Grate: Using a grater to rub food, such as vegetables, cheeses and spices, across surface to make fine pieces.

Green harvest: The harvesting of green (unripe) grapes in an attempt to increase the yield of quality grapes.

Groundnut: Peanut or monkey-nut.

Guava: A popular fruit used to make guava cheese, jelly and stewed fruit.

Gulab: Gamin Khoa or Mawa (milk evaporated to remove all moisture) is kneaded and flavoured and made into balls which are fried till golden brown and dropped immediately into hot sugar syrup.

Gustaba: Large meat balls (Kofta) in a gravy of curd, mixed with ground poppy seeds, nuts and onion, and lightly spiced.

Haggis: A kind of bag pudding, made in Scotland, from liver, sheep's head, etc., finely minced and mixed with oatmeal, herbs, etc.

Halzva: A sweet dish made of lentils, semolina or wheat, or of vegetables such as beetroot, carrot and pumpkin, with butter, milk and sugar.

Hard: A tasting term for a wine that containins too much tannin and is therefore unpleasant. Hard wines often take a long time to mature.

Haricot: A type of stew. Literally, "beans".

Hash: A made-up dish of meat that has been diced or minced. Usually a way of using leftovers.

Hectare: A metric measure that equals 10,000 m^2 (2.471 acres).

Hing: Asafoetida, a strong flavouring agent. It also acts as a carminative.

Hock: Term for Rhine wines, usually used in England.

Hogshead: A wine barrel that holds approximately 239 litres (63 gallons).

Hollandaise: A rich sauce of Dutch origin made with butter or cream, egg yolks and lemon or vinegar served hot or cold with many kinds of vegetables or fish. Hors d'oeuvre Small savoury tidbits, usually cold, served as an appetizer at the beginning of a meal.

Hopper: A counterpart of the American hot cake. For making it, the batter is fermented by the addition of toddy, an alcoholic beverage. Cooked in seasoned earthenware pots. Its subtle and pleasing flavour would appeal to all palates. Eaten with stew.

Ice wine: Wine made from frozen grapes. Called *eiswein* in German.

Idli: A fermented batter of ground lentil and rice steamed in moulds. A speciality of South India.

Indienne: A dish served in the Indian manner.

Infusion: The liquid extraction derived from steeping a substance such as coffee, tea, herbs, etc., in boiling water.

Jaggery: Refined and solidified molasses.

Jalebi: A golden-coloured, crisp sweet filled with sugar syrup, formed in rings.

Jardiniere: Garnish of vegetables.

Jeroboam: A large bottle holding three litres, the equivalent of four regular wine bottles.

Jug wine: American term for inexpensive table wine.

Juice: Extracted liquid that comes from fruits, vegetables or meats.

Julienne: Food cut into long strips. A clear soup of this name contains finely-shredded vegetables.

Junket: Milk coagulated by the addition of rennet. This makes a very light and digestible sweet. It can be served with fruit, and is a valuable nursery food.

Kababs: A savoury barbecue done on iron rods.

Kalia: A Bengali vegetable or fish dish with spices and yoghurt.

Kebab: Small pieces of meat fixed on a skewer, braised or curried.

Kedgeree: Indian dish of cooked fish, often salted or smoked, with rice, eggs etc. Sometimes curried.

Kesari: A South Indian sweet dish made of vermicelli or semolina.

Kheema: Minced meat.

Kheer: A pudding-like preparation of milk with other ingredients such as rice or carrot.

Khichdi: Mixture of rice and lentils.

Khoa Milk is boiled down till all moisture is removed, used as a base for a large number of Indian sweets.

Khorma: Rich, thickened brown curry of chicken, mutton or vegetables. Poppy seeds and desiccated coconut are added along with other condiments for thickening. The meat itself is made tender by marinading in curd (yoghurt) before cooking.

Kirsch: A favourite Continental drink is cafe kirsch; this is made with half a cup of black coffee and half a cup of cherry brandy, also a cherry cordial.

Knead: To work a dough lightly by bringing the outside of the dough into the centre, using the knuckles of the hand.

Kofta: Minced meat ball.

Kosher wine: Wine that is produced under the supervision of a rabbi so as to be ritually pure or clean. Although commonly sweet, it need not be so.

Kromeskils: Croquettes dipped in a yeast batter and fried.

La liable: The French way of saying "devilled", in other words, any very hot or highly seasoned dish.

Laddoo: A sweet ball made of gram or lentil flour, rice or semolina.

Langoustine: A small variety of spiny lobster.

Larding: To place strips of fatty substance such as fat bacon on top of lean meat or on the breast of a dry bird. To run strips of fat bacon through lean meat with a larding needle.

Lassi: A beverage made of beaten yoghurt diluted with water; sugar or salt may be added.

Late harvest wine: Also known as late picked, wine made from grapes that have been left on the vine longer than usual. Usually an indicator for a very sweet or dessert wine.

Lees: Wine sediment that occurs during and after fermentation, and consists of dead yeast, grape seeds, and other solids. Wine is separated from the lees by *racking*.

Legs: The tracks of liquid that cling to the sides of a glass after the contents have been swirled. Often said to be related to the alcohol or glycerol content of a wine. Also called *tears*.

Liaison: Mixture of eggs and cream used for thickening white sauces and soups.

Lightstruck: A tasting term for a wine that has had long exposure to Ultraviolet light causing "*wet cardboard*" type aroma and flavour.

Lime: A smaller variety of lemon.

Litchi: A luscious Indian fruit. It looks like a bigger version of the strawberry and grows in Bengal, Dehra Dun and Delhi regions.

Look: A tasting term for the casual sensory evaluation of a wine.

Macaroni: An Italian pasta made of flour forced through a tube.

Macaroons: Small cakes made from almond paste, coconut, etc.

Macedoine: A mixture, usually consisting of vegetables or fruit.

Maceration: The contact of grape skins with the must during fermentation, extracting phenolic compounds including tannins, anthocyanins, and aroma.

Macherjhal: A special Bengali fish curry where the fish is first fried and then curried. Mustard oil is the cooking medium used. Mackerels make excellent jhal.

Machhi: Fish.

Madeirized: A wine showing Madeira-like flavour, generally evidence of oxidation. Sometimes used to describe white wine that has been kept long past its prime.

Magnum: A bottle holding 1.5 litres, the equivalent of two regular wine bottles.

Maida: Refined flour.

Maize: Indian corn.

Majoram: A lemon-scented herb.

Maltre d'hotel butter: A piquant sauce or butter flavoured with parsley and lemon.

Mango: A very luscious Indian fruit, which has several varieties-- Dasehri, Chausa, Langara, Alfonso, etc.

Marinade: A process of flavoring food by soaking in a liquid or dry mixture.

Marinate: To steep in a mixture of yoghurt or vinegar to soften.

Marzipan: A sweet or icing made of almonds ground to a paste with egg.

Masala: Indian condiments and spices.

Mash: Removing all lumps from food with either a fork or electric beater.

Mask: To coat a dish with sauce or line a mould with jelly.

Matignon: A mixture of finely minced carrot, celery, onion, ham, thyme and crushed bay leaf used to cover joints or poultry before being roasted or cooked to impart a good flavour.

Mattar: Peas.

May wine: A light German wine flavored with sweet woodruff in addition to strawberries or other fruit.

Mayonnaise: A thick sauce made of egg yolks, oil and vinegar, mustard, etc. Used as a dressing for salads of all kinds.

Melt: To heat a solid food until it becomes liquid or semi-liquid.

Menu: List of fare.

Meringue: A mixture of stiffly beaten egg white and sugar, often used as a garnish when browned in the oven. It can be made into small sugar cakes and afterwards filled with cream.

Méthode Champenoise: Process whereby sparkling wines receive a second fermentation in the same bottle that will be sold to a retail buyer. Compare with Charmat or bulk fermented.

Methuselah: A large bottle holding six litres, the equivalent of eight regular wine bottles.

Mignonette pepper: White peppercorns, coarsely ground.

Millerandage: A French term referring to a viticultural problem in which grape bunches contain berries of greatly differing size and levels of maturity. Caused by cool weather during flowering.

Minestrone: Italian soup with many vegetables and macaroni added.

Mirepoix: Roughly cut onions, carrots, celery and a sprig of thyme and bay leaf.

Mise-en-place: A general name given to the elementary preparation area which is constantly resorted to during the various stages of most culinary operations.

Mix: Beating or stirring foods together until they are incorporated.

Mocha: A flavouring usually made with a coffee infusion.

Moilee: Fish or prawn cooked in plenty of coconut milk--lightly spiced.

Mousse: A light spongy dish, made with sweetened and flavoured cream, then whipped .and frozen. Can also be made with a mixture of meat, vegetables, etc., mixed with gelatine.

Mousseline: Same as above but moulded in small quantities. Enough for one person at a time.

Much sharper: Loochi Bengali version of puri.

Muffin: A drop batter baked in small individual moulds and eaten hot with butter.

Mulled wine: Wine that is spiced, heated, and seved as a punch.

Mulligatawny Curry: Chicken or mutton curry with a flavour popularised in the West as Mulligatawny soup.

Murgh Musallam: A spicy chicken preparation.

Must weight: The level of fermentable sugars in the must and the resultant alcohol content if all the sugar was converted to ethanol.

Must: Unfermented grape juice, including pips, skins and stalks.

Mutanjan: Spiced rice with mutton sweetened with sugar. It is a speciality of Kashmir.

Nan: Indian bread made of slightly leavened dough, baked on the wall of a mud oven.

Nargisi Kofta Curry: Minced meat, mixed with spices and egg, covers a hardboiled egg (like scotch eggs) and it is fried and dropped in a spiced curry.

Nebuchadnezzar: A large bottle holding 15 litres, the equivalent of 20 regular wine bottles.

Neera: A non-fermented drink obtained from the coconut palm.

Négociant: French for "*trader*". A wine merchant who assembles the produce of smaller growers and winemakers and sells the result under its own name.

New World wine: Wines produced outside of the traditional wine growing areas of Europe and North Africa.

Noble rot: Another name for the *Botrytis cinerea* mould that can pierce grape skins causing dehydration. The resulting grapes produce a highly prized sweet wine, generally dessert wine.

Noodles: A flour pasta served in small fancy shapes or in thin string-like tubes. Often fried in Chinese fashion.

Nose: A tasting term for the aroma or bouquet of a wine.

Nougat: A sweet of a fairly rich kind made with almonds, sugar, nuts, cherries and honey.

Oak chips: Small pieces of oak wood used in place of oak barrels in fermenting and/or ageing wine.

Oenology: The science of wine and winemaking.

Oenophile: A wine aficionado or connoisseur.

Old vine: Wine produced from vines that are notably old.

Old World wine: Wines produced inside of the traditional wine growing areas of Europe and North Africa.

Pachadi: Seasoned yoghurt with vegetables--served in South India.

Palak Mutton or Palak Chicken: Meat or chicken cooked with spinach, lightly spiced and thick, without much gravy. The spinach is mashed and cooked with the meat.

Palate: A tasting term for the feel and taste of a wine in the mouth.

Panada: A thick paste of flour and liquid or flour and butter with a little liquid used to bind together ingredients which would fall apart by themselves. Paneer to coat in egg and breadcrumbs.

Pani puri: A puff of dough fried and eaten with tamarind water and sprouted gram.

Panir: Cottage cheese made by curdling milk with lemon juice, curd etc., and the whey removed by straining and hanging in a muslin bag.

Papaya: A luscious Indian fruit; papaw or pawpaw.

Papillate: Paper frills placed on chops for decorations. It also means food cooked in paper casings.

Paratha: Fried bread made of unleavened wheat flour.

Parboil: To boil food until only partly cooked.

Pare: To peel. Parfait Ice-cream served in a tall glass and decorated with a variety of nuts and fruits.

Parmesan: A very hard cheese made in Italy from cow's milk. Used mainly for cooking.

Pate: Pie, pastry or raised pie, or a paste.

Paysanne: To cut into even, thin pieces, triangular or round or square.

pH: An acronym for "*potential hydrogen*", in wine the term refers to the level of acidity. The lower the pH, the higher the acidity.

Phirnee: A creamy milk pudding made of rice flour. Pinafore a type of white fish very popular in Bombay.

Phylloxera: A microscopic underground insect that kills grape vines by attacking their roots.

Pimento: Red or green pepper pods used in salads or often as a colourful garnish.

Pip: Grape seeds.

Piquante: Sharply flavoured. Usually the word is applied to mustard or a sharp sauce.

Pistachio: Green-coloured nut kernels. Used as a garnish.

Pit: Taking the seed out of a piece of fruit or vegetable.

Plasma: A South Indian milk pudding made from cereal (such as rice or vermicelli) or pulse, and sugar.

Plat do jour: Special dish of the day.

Plonk: British English slang for an inexpensive bottle of wine. The term is thought to originate from the French word for white wine, "*blanc*".

Pluck: To remove feathers from poultry and birds.

Poach: Cooking food by placing in simmering liquid.

Poach: To cook just below boiling point in hot liquid in an open pan.

Poele: Roast done entirely or almost entirely in butter. It is a simplified process of old cookery. The joint or poultry is first fried and coated with a thick layer of matignon, wrapped in slices of pork fat covered with butter paper or aluminium foil and cooked in the oven. It is basted with melted butter during cooking.

Pomace: The skins, stalks, and seeds that remain after making wine. Also called *marc*.

Potage: A nourishing broth or soup.

Pot-au-feu: A beef soup made with vegetables, and poured over French bread or toast.

Potpourri: A stew of various meats and spices. Praline Burnt almond flavouring.

Prawn malai curry: Prawns cooked in spiced cream and coconut cream and baked in a coconut.

Pre-heat: Heating the oven to the temperature needed before using it.

Proof: Allowing a yeast dough to rise before baking.

Proof: Refers to the alcohol content of a beverage. In the United States, proof represents twice the alcohol content as a percentage of volume. Thus, a 100 proof beverage is 50% alcohol by volume and a 150 proof beverage is 75% alcohol. In the Imperial system, proof, (or 100% proof), equals 57.06% ethanol by volume, or 48.24% by weight. Absolute or pure ethanol is 75.25 over proof, or 175.25 proof.

Pulao rice: A special type of rice used in pulao -- thin and long and not as starchy as ordinary rice.

Pulao: A rice dish. Rice is fried in fat and cooked in stock and water.

Pulses: Vegetables that grow in pods, e.g., peas, beans, lentils, etc.

Pumpkin: A big, round vegetable of the squash variety.

Puncheon: A wine barrel that holds approximently 318 litres (160 U.S. gallons).

Punt: The indentation found in the base of a wine bottle. Punt depth is often thought to be related to wine quality, with better quality wines having a deeper punt.

Puree: A smooth mixture obtained by rubbing cooked fruit, vegetables, etc., through a sieve.

Puree: Blending a food into a liquid or heavy paste.

Puri Deep-fried round wheat bread: About 5 cm. or 2 inches in diameter and puffed,

Qualitätswein mit Pradikat: A designation of best quality German wines that must conform to specific requirements of origin and composition.

Qualitätswein: A designation of better quality German wines.

Quebracho: A type of meat kabab.

Quenelles: Forcemeat or meat, fish, game or poultry, pounded, rubbed through a sieve and formed into balls, then poached or fried.

Rabri: Sweet, thickened milk.

Racking: The process of drawing wine off the sediment, such as lees, after fermentation and moving it into another vessel.

Ragout: Well-flavoured meat stew. Thick, well-seasoned and rich.

Raising agents: Substances which produce a gas when

acted on by heat or other substances, and make flour mixtures rise, e.g. baking powder, baking soda, yeast.

Raita: Vegetable, raw or parboiled, mixed with beaten and seasoned yoghurt.

Ramekins: Individual small baking dishes.

Rasam: A spicy soup made of lentils and tomatoes. Hot and sharp; served sometimes as an appetiser. A speciality of South India.

Rasgullas: Milk is curdled by the addition of curd and lime, the casein separated from whey which is then kneaded smooth and made into balls. A piece of sugar candy is usually kept in the centre. These balls are put into clarified simmering sugar syrup and cooked.

Raspings: Very fine breadcrumbs obtained by grating stale or over,-dried bread on a fine grater.

Ratafia: A flavouring. Tiny biscuits for trifles, etc., flavoured with almonds.

Rava: Semolina.

Ravioli: A savoury meat mixture enclosed in dough and cooked in boiling water.

Rechauffe: A reheated dish.

Reconstitute: Bringing condensed or concentrated food to its original strength by adding water.

Reduce: Briskly boil liquids, so that the liquid evaporates to thicken the sauce making it rich and flavorful.

Rehoboam: A large bottle holding 4.5 litres, the equivalent of six regular wine bottles.

Relish: A highly-seasoned food used as an accompaniment.

Reserva: Spanish and Portuguese term for a reserve wine.

Reshmi chapatties: Unleavened bread resembling tortillas, but as thin as silk, and about 20 to 25 cm. or 8 to 10 inches in diameter. Baked over an upturned domelike iron vessel over live coal.

Residual sugar: Also known as RS, the level of sugar that remains unfermented in a wine. *See also sweetness of wine.*

Reverse osmosis: A process used to remove excess alcohol from wine made from intentionally overripe grapes.

Rex: A thickening for soups or sauces made with flour and fat.

Riddling: Also known as *"Rémuage"* in French, part of the Méthode Champenoise process whereby bottles of sparkling wine are successively turned and gradually tilted upside down so that sediment settles into the necks of the bottles in preparation for *degorgement*. Part of the Méthode Champenoise process.

Rista: A minced meat ball curry, a speciality of Kashmir

Roast: Cooking an uncovered piece of meat, poultry, o vegetables in an oven.

Roe: Eggs offish.

Rogan josh: A red curry with a thin gravy made from the leg of mutton cut with bones into fairly large pieces. The distinct flavour is that of saffron and nutmeg. Saffron is added just before removing from the fire. The red colour is obtained by the addition of saltpetre and rattanjog, a red coloured bark.

Rosé wines: Pink wines that are produced from the shortened contact of red wine juice with its skins, reducing the red colour of the wine. These wines can also be made by blending a small amount of red wine with white wine.

Royal: The name of a kind of icing.

Royale: A savoury egg custard cut into cubes and used as garnish for a soup.

Ruby: A style of Port that is generally sweet.

Sabayon: Yolks of eggs and a little water cooked until creamy.

Sack: An early English term for what is now called Sherry.

Saffron: The dried stigma of the crocus flower, used for adding colour and flavour.

Salmanazar: A large bottle holding nine litres, the equivalent of 12 regular wine bottles.

Salmi: Partly roasted game made into a rich stew.

Sambar: A preparation of lentils with one or more types of vegetables-- served as a necessary accompaniment to rice in South Indian vegetarian food.

Samosa: A thin pastry cone filled with boiled or spiced vegetable or minced meat and deep fried.

Sandesh: Sweetened milk, condensed to solid form by slow cooking and flavoured.

Sangria: A tart punch made from red wine along with orange, lemon and apricot juice with added sugar.

Sauce: A pouring mixture, sweet or savoury, to serve with, and enhance the flavour of another dish.

Saute: Cooking on a stove in a small amount of oil or butter until tender.

Scald: To immerse food in boiling liquid for a short time. To heat a liquid (e.g. milk) to just under boiling point or to pour boiling water on (e.g. scald a jug).

Scallop: Food baked in layers with sauce and breadcrumbs.

Score: To cut slits through the outer surface of food before cooking to tenderize or to make a decorative pattern.

Screwcap: An alternative to cork for sealing wine bottles, comprising a metal cap that screws onto threads on the neck of a bottle. Also called a "Stelvin".

Sear: Quickly browning meat on all sides with high heat to seal in the juices.

Seasoned flour: Flour flavoured with salt and pepper.

See: A fried savoury resembling vermicelli. It is made of gram flour dough passed through a mould and deep fried till crisp.

Seekh kabab: Minced meat and spices ground together bound with egg and fixed on to a skewer and grilled.

Semi-generic: Wines made in the United States but named after places that the Alcohol and Tobacco Tax and Trade Bureau requires be modified by a US name of geographic origin. Examples would be New York Chablis, Napa Valley Burgundy or California Champagne.

Shami kabab: Cutlets made with minced meat and lentil.

Sherbet: Frozen mixture of fruit juice, egg whites, sugar, milk or water.

Sherry: A fortified wine that has been subjected to controlled oxidation to produce a distinctive flavor.

Shortening: Fat suitable for baking is sometimes called this.

Shred: Making long narrow strips of food with a food processor or a grater.

Shredded: Cut in fine strips e.g., lettuce, cabbage, etc.

Shrikhand: A sweet yoghurt based dish with liquid removed by straining, and flavoured with saffron and nuts.

Sift: Putting dry ingredients such as flour or sugar through a sifter or mesh screen to loosen particles and incorporate air.

Sift: To put dry ingredients through a fine sieve.

Simmer: A method of cooking food in liquid that is kept just below the boiling point.

Skewer: A long pin of wood or metal used to secure meat or poultry while cooking.

Skim: Removing any fat or foam from the surface of liquid.

Snake gourd: Vegetable with a long snakelike shape.

Solera system: A process used to systematically blend various vintages of Sherry.

Sommelier: A trained wine expert that often works in fine restaurants.

Sorbet: A half-frozen water ice, served in the middle of a long dinner.

Souffle: Very light baked or steamed pudding, savoury or sweet, usually puffed up with egg whites.

Souse: Fish such as herrings, pickled in vinegar and spices.

Spaghetti: An Italian pasta, finer than macaroni, coarser than vermicule.

Sparkling wine: Effervescent wine containing significant levels of carbon dioxide.

Spätlese: German for "*select*". Generally applied to German late harvest wines.

Split: A wine bottle that holds 375 mL, half the equivalent of a typical 750 mL bottle.

Spumante: Italian for "*sparkling*". Generally any sparkling wines from Italy.

Steam: A method of cooking food in the vapor given off by boiling water.

Steep: Letting a food, such as tea, stand in water that is just below boiling to extract flavor and color.

Stew: To cook by simmering in a little liquid.

Still wine: Wine that is not sparkling wine.

Stir: Incorporating ingredients with a spoon to prevent them from sticking during cooking or to cool them after cooking.

Stock: Well-flavoured liquid made from meat, vegetables, fish or poultry, etc., and used as a foundation for soups, sauces, stews, etc.

Sulfites: Compounds added to wine to prevent oxidation and microbial spoilage.

Sulphur dioxide: A substance used in winemaking as a preservative.

Supreme: A name given to the fillet of fish or breast of fowl or game.

Sweetness of wine: Defined by the level of residual sugar in the final liquid after the fermentation has ceased. However, how sweet the wine will actually taste is also controlled by factors such as the acidity and alcohol levels, the amount of tannin present, and whether the wine is sparkling.

T budding: A technique that permits grafting of different grape varieties onto existing rootstocks in a vineyard.

Table d'hote: A set meal at a fixed price.

Table wine: Generally any wine that is not sparkling or fortified. In the US these wines must also be between 7% and 14% alcohol by volume.

Tamarind: A sour fruit used for curries, chutneys, etc.

Tandoori Chicken or Fish: Spiced meat or fish barbecued in a specially designed tender oven.

Tannin: Polyphenolic compounds that give wine a bitter, dry, or puckery feeling in the mouth.

Tart: A tasting term describing a wine high in acidity. Often displayed by young, unripe wines.

Tartare: A cold, sharp, savoury sauce with a base of mayonnaise served with fried fish or a meat dish.

Tartaric acid: The most important acid found in grapes.

Tasting flight: Refers to a selection of wines, usually between three and eight glasses, but sometimes as many as fifty, presented for the purpose of sampling and comparison.

Tepid: The temperature of a mixture of 2 parts of cold water to 1 part of boiling water, i.e. about 26.6° or 80°F

Terrine: A baked, savoury meat or game mould which is cooked in the oven, usually in an earthenware dish.

Terroir: French for "*soil*", the physical and geographical characteristics of a particular vineyard site that give the resultant wine its unique properties.

Texture: A tasting term for the mouthfeel of wine on the palate.

Thief: A tubular instrument for removing a sample from a cask or barrel. Also called a *pipe*.

Timbale: A cup-shaped mould, usually made from meat, fish or vegetables, added to a custard mixture, and finely decorated.

Toast: Heating nuts, seeds, coconut or bread in the oven until they are slightly browned. This will bring out their natural flavor.

Toddy: An alcoholic drink obtained from a variety of palm.

Tournedos: Small fillet steak cut from the middle of tenderloin (nearly 65 mm. (1/4") thick and about 6-7 cm. (2½ to 21/3") in diameter.

Tranche: A slice.

Trangon: A slice of flat fish on the bone.

Truss: To tie up or skewer a bird ready for the oven after plucking and drawing.

Tun: A wine cask that holds approximately, two butts, or 252 U.S. gallons.

Turmeric: A yellow spice used for imparting a characteristic flavour and colour. The root of a plant belonging to the ginger family.

Tutti-frutti: Mixed fruit.

Ullage: Also known as headspace, the unfilled space in a wine bottle, barrel, or tank.

Unoaked: Also known as unwooded, refers to wines that have been matured without contact with wood/oak such as in aging barrels.

Uppuma: A savoury South Indian dish made out of semolina and lentils.

Vanilla: A flavouring generally used for cakes, ices or puddings.

Varietal: Wines made from a single grape variety.

Veloutes: A rich, creamy smooth sauce.

Vermicelli: A very fine Italian Pasta.

Vermouth: A fortified wine that has been flavoured with as many as 40 herbs and spices.

Vertical and horizontal wine tasting: In a vertical tasting, different vintages of the same wine type from the same winery are tasted. This emphasizes differences between various vintages. In a horizontal tasting, the wines are all from the same vintage but are from different wineries. Keeping wine variety or type and wine region the same helps emphasize differences in winery styles.

Vindaloo: A sharp and hot curry made of pork, prawn, chicken or mutton. Originated in Goa.

Vine: A plant on which grapes grow.

Vinegar: A sour-tasting, highly acidic, liquid made from the oxidation of ethanol in wine, cider, beer, fermented fruit juice, or nearly any other liquid containing alcohol.

Vineyard: A place where grape vines are grown for wine making purposes.

Vinho verde: An effervescent white wine produced in Portugal.

Vinho: Portuguese for *wine*.

Viniculture: The art and science of making wine. Also called enology (or oenology). Not to be confused with viticulture.

Vinification: The process of making grape juice into wine.

Vintage: The year in which a particular wine's grapes were harvested. When a vintage year is indicated on a label, it signifies that all the grapes used to make the wine in the bottle were harvested in that year.

Viticulture: The cultivation of grapes. Not to be confused with viniculture.

Vol-au-vent: Creamed game, meat or fish served in a pastry case.

Wada or Pakora: A rissole of lentil or vegetable served as a snack.

Walewska: Garnish for fish or collops of lobster, slices of truffles.

Wark: Thin foil of silver or gold which is edible.

Whip: Beating food rapidly by using a whisk or electric beater to incorporate air into the mixture.

Whisk: To beat cream or eggs until a stiff froth is obtained.

Wine cave: A large cave that is excavated to provide a cool location for storing and aging wine. Similar to wine cellar.

Wine cellar: A cool, dark location in which wine is stored, often for the purpose of ageing.

Wine fault: Undesirable characteristics in wine caused by poor winemaking techniques or storage conditions.

Wine label: The descriptive sticker or signage adhered to the side of a wine bottle.

Wine lake: Refers to the continuing surplus of wine over demand (glut) being produced in the European Union.

Wine tasting: The sensory evaluation of wine, encompassing more than taste, but also mouthfeel, aroma, and colour.

Winemaker: A person engaged in the occupation of making wine.

Wine-press: A device, comprising two vats or receptacles, one for trodding and bruising grapes, and the other for collecting the juice.

Winery: A building, property, or company that is involved in the production of wine.

Yeast: A microscopic unicellular fungi responsible for the conversion of sugars in must to alcohol. This process is known as alcoholic fermentation.

Yellow rice: Rice flavoured and coloured with saffron or turmeric.

Young: Wine that is not matured and usually bottled and sold within a year of its vintage.

Zarda: A sweet pulao served at the end of a meal.

Zest: Thin outer skin of oranges and lemons.

Zymology: The science of fermentation.

Bibliography

Arnaldo, M.J (1981), "Hotel general managers: a profile", *The Cornell Hotel and Restaurant Administrative Quarterly*, Vol. 22 No.3, pp. 53-6.

Arnold, J, Robertson, I.T, Cooper, G.L (1991), "Career choice and development", *Work Psychology: Understanding Human Behaviour in the Workplace*, Pitman, London, pp. 265-82.

Bentivenger, A, Sluder, L (1989), "Hotel GMs bring diverse background to the job", *Lodging*, Vol. 14 No.10, pp. 61-2.

Brownell, J. (1992), "Personality and career development", *The Cornell Hotel and Restaurant Administrative Quarterly*, Vol. 35 No.2, pp. 36-43.

Carper, J (1993), "The painful truth: operations is no longer King", *Hotels*, Vol. 50.

Chartrand, J.M, Camp, C.C (1991), "Advancements in the measurement of career development constructs: a 20-year construct", *Journal of Vocational Behaviour*, Vol. 39 No.2, pp. 1-39.

Clark, M (1993), "Communications and social skills: perceptions of hospitality managers", *Employee Relations*, Vol. 15 No.2, pp. 51-60.

Cooper, C., Westlake, J. (1998), "Stakeholders and tourism education", *Industry and Higher Education*.

Cracknell, H.L., Kaufmann R.J., Nobis, G. (1983), *Practical Professional Catering*, Macmillan, Basingstoke, pp. 22.

Dex, S (1991), *Life and Work History Analyses: Qualitative and Quantitative Developments*, Routledge, London.

Doswell, R., Gamble, P.R. (1979), *Marketing and Planning Hotels and Tourism Projects*, Hutchinson, London, pp. 19-20.

Gamble, P.R, Lockwood, A, Messenger, S (1994), "European management skills in the hospitality industry", University of Surrey.

Guerrier, Y, Lockwood, A (1989), "Developing hotel managers - a reappraisal", *International Journal of Hospitality Management*, Vol. 8 No.2, pp. 82-9.

Guerrier, Y, Lockwood, A (1990), "Managers in hospitality: a review of current research", *Progress in Tourism, Recreation and Hospitality Research*, Vol. 2 pp. 151-67.

Guerrier, Y. (1987), "Hotel managers' careers and their impact on hotels in Britain", *International Journal of Hospitality Management*, Vol. 6 No.3, pp. 121-30.

Harris, N. (1989), *Service Operations Management*, Cassell, London, pp. 3.

Hay, J (1990), "Managerial competencies or managerial characteristics", *Journal of Management Education and Development*, Vol. 21 No.5, pp. 305-15.

Haywood, K.M (1987), "Thoughts on thinking: a critical human resource skill", *The Cornell Hotel and Restaurant Administrative Quarterly*, Vol. 28 No.2, pp. 51-5.

Hicks, H.G., Gullett, C.R. (1976), *The Management of Organisations*, 3rd International Studen Edition, McGraw-Hill Kogakusha Ltd, Tokyo, pp. 13.

Hicks, L (1990), "Excluded women: how can this happen in the hotel world?", *Service Industries Journal*, Vol. 10 No.2, pp. 348-63.

Hill, T. (1983), *Production / Operations Management*, Prentice-Hall, Englewood Cliffs, NJ.

Howard, B. (1992), *Operations Management*, Macmillan, Basingstoke, pp. 1.

Hsu, J.F, Gregory, S (1995), "Developing future hotel managers in Taiwan: from an industry viewpoint", *International Journal of Hospitality Management*, Vol. 14 No.3/4, pp. 261-9.

Ladkin, A, Riley, M (1996), "Mobility and structure in the career paths of UK hotel general managers: a labour market hybrid of the bureaucratic model?", *Tourism Management*, Vol. 17 No.6, pp. 443-52.

Lee, S.M., Schniederjans, M.J. (1994), *Operations Management*, Houghton Mifflin Company, Boston, MA, pp. 3.

Messenger, S. (1991), "The UK hospitality and tourism industry: an overview of the issues affecting the supply of education and training in the 1990s", *Progress in Tourism, Recreation and Hospitality Management*, Vol. 3 pp. 247-63.

Nebel, E.C.III, Braunlich, C.G, Zhang, Y (1994), "Career paths in American luxury hotels: hotel food and beverage directors", *International Journal of Contemporary Hospitality Management*, Vol. 6 No.6, pp. 3-9.

Nightingale, M. (1985), " The hospitality industry: defining quality for a quality assurance programme - a study in perceptions", *Service Industries Journal*, Vol. 5 No.1, pp. 9-22.

O'Driscoll, M.P, Humphries, M, Larsen, H.H (1991), "Managerial activities, competencies and effectiveness: manager and subordinate perceptions", *International Journal of Human Resources Management*, Vol. 3 No.2, pp. 313-26.

Pavesic, D.V (1993), "Hospitality education 2005: curricular and programmatic trends", *Hospitality Research Journal*, Vol. 17 No.1, pp. 285-94.

Ruddy, J (1990), "Patterns of hotel management development in South East Asia", *Hospitality Research Journal*, Vol. 14 No.2, pp. 349-61.

Ruddy, J. (1989), "Career development of hotel managers in Hong Kong", *International Journal of Hospitality Management*, Vol. 8 No.3, pp. 215-25.

Slack, N. (1983), "Operations management and curriculum design", *Management Education and Development*, Vol. 14 No.1, pp. 19-32.

Tarpey, D (1991), "Mind your language", *Caterer and Hotelkeeper*, Vol. 184 No.3691, pp. 40-42.

Index

■■■